SUSTAINABLE KITCHEN

SUSTAINABLE KITCHEN

Projects, tips and advice to shop, cook and eat in a more eco-conscious way

ABI ASPEN GLENCROSS
& SADHBH MOORE

Contents

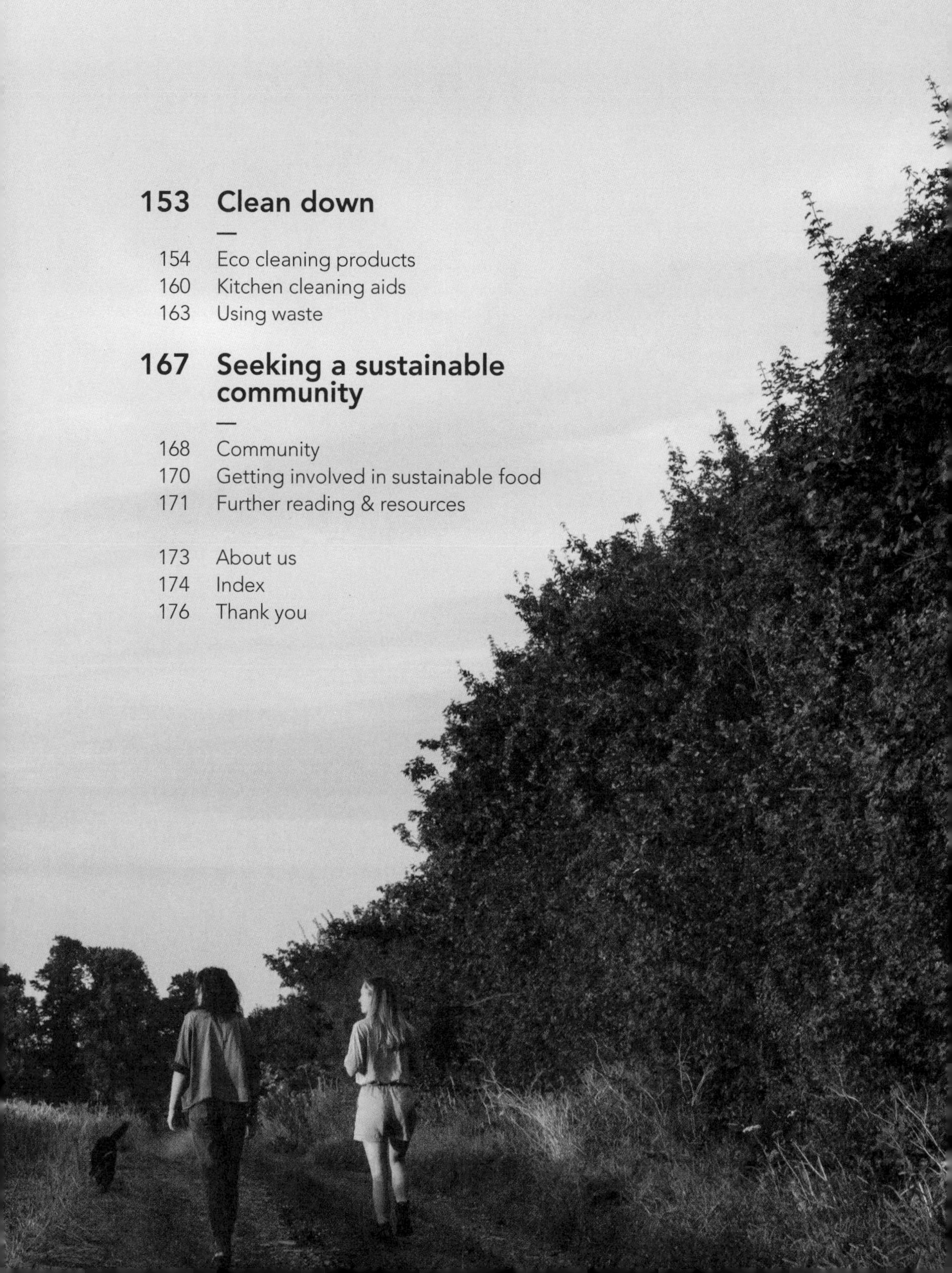

Sustainable food

We wanted to start with a little insight into our sustainability considerations for many of the things that a kitchen exists to process – food ingredients. Each food group we've listed, or even ingredient, could have a whole book written about it; the myriad complex, nitty gritty things to look into when we really dissect each food and its sourcing, and how it links to environmental sustainability. We realise not everyone has the time, or desire, to go into that much detail. We have tried to capture some of the nuance and complexity of issues of food sustainability, but at a surface level, allowing readers to look into each topic in greater detail if they wish. Further reading suggestions and sources of information on all topics food sustainability are mentioned throughout the book (see pages 168 and 171).

Interpreting sustainable

We embarked on our journeys into food and farming after discovering that the food and agriculture sector is one of the biggest greenhouse gas emitters; it is estimated that up to one-third of all of agriculture's emissions are attributed to animal production. If we want to live more sustainably we realized that we needed to start in our kitchens.

The term sustainable, in the context of this book, refers primarily to the environmental branch of sustainability, whereas sustainability in reality often refers to a balance between social, economic and environmental priorities. We hope it becomes clear throughout the book that while there may not necessarily be a definitive 'sustainable kitchen' model, there are lots of actions you can take and considerations you can make in moving towards a more sustainable kitchen.

Of course, honing a more sustainable kitchen cannot alone 'save the planet', but it can at least empower. Collective individual actions add up and can help to reach the tipping point needed for measurable change. But maybe more importantly, it can help reduce a feeling of helplessness, and sometimes guilt, that many of us may have experienced. We hope to furnish readers with a new hobby, ultimately. Some of these guides, practices and recipes can hopefully become part of your day-to-day, and can help in the transition towards a more hopeful, inspired, inquisitive approach to environmental sustainability, food and the kitchen.

Striving for sustainability is a complex matter. Sometimes the ingredients, recipes or kitchen hacks mentioned will be focused on reducing environmental impact, and that choice may also be the optimum for social sustainability too – using Fairtrade organic cocoa, for example. This may not always seem to be the most economically sustainable option for consumers, because Fairtrade cocoa can cost more than non-Fairtrade cocoa. Other times, by prioritizing environmental sustainability, it may also be the most economically sustainable option for us too – buying secondhand or refurbished electronics, for example. This may not always feel like the most socially sustainable option – will buying less mean more businesses don't survive?

Sustainable food was a hot topic, but there weren't many opportunities to experience what a sustainable food system be like. We wanted to help bring it to the table and make it tangible, or technically, gustable, or tasteable. We saw that we could facilitate the intersection of science, activism and the food system, so began hosting food events. These events not only told the stories of food sustainability and suppliers, but also literally served up what it might look and taste like to diners.

For us at The Sustainable Food Story, running our kitchens and events with environmental sustainability at their core has had a dual effect of contributing to our economic sustainability.

We are minimalists. We aim to be frugal where needed. We focus on being resourceful and innovative. We are not very glamorous or flashy, but try to make up for that in creativity and detail. Although it may seem like aiming for a more sustainable kitchen will be an expensive endeavour, we hope that there is a balance in higher costs (perhaps of certain ingredients) versus money saving (shopping seasonally or secondhand) and expenses should level out.

We recognize this unique position in focusing so much time, effort and financial resources on prioritizing the environmental impacts of our food choices and kitchen sustainability. We want to share as much as we can from all of the time we've spent obsessing over minor details and

falling down rabbit holes of sustainability as well as learning from the mistakes that can be made. By being so obsessive about the details of these subjects we have been fortunate enough to turn this passion and hobby into a career.

Much environmental literature and commentary highlights that there will have to be compromise and sacrifice. We prefer to see this as a challenge rather than a restriction. We imagine that many can relate to sometimes feeling overwhelmed and confused by having too many options.

By forming your own sustainable kitchen rationale and guidelines based on some of the information in this book we hope that you can enjoy the challenge of doing things differently. We hope you can find some new approaches to some of the more routine and normalized ways of running a kitchen. Whether that is having vegetables as the focus of your meal instead of red meat, or minimizing how often you buy kitchen towel, we hope there's something for everyone between these FSC-certified pages.

Impact

We have long grappled with the question, 'Should individuals be taking on the burden of responsibility to live more sustainably?' We are fully aware of, and outspoken about, the fact that there needs to be systemic changes to many sectors and the 'business-as-usual' approach, in order to reduce global carbon emissions. Individuals cannot bear the full brunt of changing their lives to try to keep global temperature rise down (they are now an average 1.5°C/34.7°F warmer than pre-industrial times). We have come to the conclusion, as have many others, that we need both a top-down and a bottom-up approach. We need strong governmental leadership and appropriate policies, industry accountability, legally binding global agreements and technological assistance if we want to stand any chance of reducing the impacts of the climate crisis. But we also need grassroots community movements, individual behaviour change and actions to adapt our lifestyles and our homes. And we need to keep pressure on, to drive change from the bottom up. Much has been written on the various scenarios, theories and approaches to climate change mitigation, adaptation and environmental conservation. There are many books specifically on the impact of agriculture and our food systems on our planet, and subsequently on us and future generations. Although this responsibility can become overwhelming, it's how we made a lifelong connection the first night we met. Cooking food, chatting waste and bonding over our shared reverence for Dan Barber and excitement over his book *The Third Plate*.

How we eat: our food philosopy

When it comes to our food philosophy, we started with the low-hanging fruit, so to speak, making food choices based on the scale of the adverse contribution of those foods to the planet and its ecosystems. Having vegetables, whole grains and pulses as the majority of our plate leaves little room, or need, for animal-derived foods, which tend to have a higher environmental burden. When we do eat animal products, they tend to serve as a garnish or seasoning. We make substitutions, and use alternative ingredients, not sticking rigidly to recipes or traditional meal structures like the old 'meat and two veg' plate. We could be considered flexitarians, or reducetarians, if anyone really wanted to put a label on our relationship with meat-eating.

In our kitchen, meals are focused around seasonal vegetables. Buying ingredients in season means they're likely to be at their tastiest, locally grown and have a lower environmental impact. We came to realize early on in our exploration of food sustainability that diversity is key, for us and the planet. We fill our plates with a range of colours, textures and varieties of all the food types we eat. We try to plan our meals, incorporating leftovers, and using as much of the plant as we can, composting anything we can't use and reducing waste.

We prioritize buying direct from farmers or trusted suppliers, such as at the farmers' market, or ideally grow our own. This ensures greater traceability and transparency, and gives us a more in-depth knowledge of the potential impacts of our food choices on the environment. We endeavour to source much of our food from local businesses who are championing sustainability, as we want to support them, again aiding with the transparency in the supply chain. This is also a great opportunity to buy unpackaged. We aim to minimize packaging, buying dry goods in bulk, refillable containers or recyclable packaging where possible.

We check labels, and minimize how much processed food we buy, preferring to buy wholefoods. Labels like organic, biodynamic, Fairtrade and MSC (Marine Stewardship Council) ensure some level of environmental and social consideration and help guide our choices.

At The Sustainable Food Story, our ultimate aim is to change the food system for the better, from field to fork to challenge the status quo, campaign and aid in creating a just, delicious and enduring food system for all.

Environmentally linked diet definitions

There are so many terms we can use to label ourselves with:

- *Vegan:* only eats plant foods, avoids animal-derived products.
- *Plant-based:* mostly eats plant foods, with plants or vegetables as the focus of the meal, but might not be as strict as vegan.
- *Vegetarian (lacto-vegetarian, ovo-vegetarian):* eats plant foods, as well as dairy and/or eggs. Nothing directly from animal slaughter.
- *Pescatarian:* eats fish and seafood but not meat.
- *Flexitarian (semi-vegetarian):* centred around plant food, with the occasional inclusion of meat.
- *Reducetarian:* reducing the amount of meat eaten.
- *Climavore:* eats primarily for the climate, in reaction to climatic events and conservation challenges.
- *Locavore:* eats food grown and produced within a certain radius to them.
- *Omnivore:* eats plants and animals.

Our rationale

We don't really label ourselves as this often puts boundaries or limits on what we eat or our thought processes, which are ever-evolving with research. We don't really fit into one category, or maybe we fit into them all at different times? Essentially we eat a veg-centric diet with lots of whole grains, pulses, seasonal fruit and plenty of nuts and seeds. We eat meat in small amounts, using it to flavour dishes, and we usually choose a byproduct over prime cuts, or we will eat meat if it's leftovers or the only option. We also don't nail ourselves for having some treats, as we do love a good biscuit, but try to make our own most of the time. Our diet does change in reaction to what is happening in the world, what we learn, what our bodies need and what needs eating right now to regain balance in the world.

HOW TO SHOP SUSTAINABLY

Vegetables: prioritize farmers' markets and farm shops where it's easy to buy seasonal, local, unpackaged and often organic. Organic from farmers' markets can even be cheaper than organic from supermarkets. Support small, mixed farms.

Fruit: same as above. Rarely buy tropical fruits, – they are often air-freighted (think papaya and mango) – other than as a special treat, or when it's reduced to clear and likely to be thrown out (and often more likely to be ripe). If occasionally buying bananas or avocados, look for Fairtrade and organic. Frozen and dried fruit are good options for avoiding food spoilage, and preserve fruits when in season. We steer clear of pre-made juices because of the associated pulp waste and high sugar content without fibre.

Grains: choose whole grains and opt for grains grown in your geographical location. Look for heritage and heirloom varieties, which have often been bred for regenerative farming systems. Opt for stone-milled whole grains over white flours to reduce bran wastage and increase nutrient content. If choosing white flour, make sure it's unbleached. Be mindful of brown or wholemeal flour labels, as they are often missing the nutrient-dense germ. We like Duchess Farms and Gilchesters in the UK, and Anson Mills and Maine Grains in the US.

Pulses: easy to buy cheaply, in bulk, with minimal packaging – dried lentils, mung beans and broad beans, for example. Look for regional producers of heritage varieties (Hodmedods in the UK and Rancho Gordo in the USA as an example).

Meat: Use sparingly, not as the main part of a meal but rather as a garnish or flavouring (such as using bones to make stock, or use smaller portions than suggested in standard recipes. Opt for blood pudding or offal if looking for an iron boost – more nutritious and underused, so doesn't drive demand as prime cuts like steak would. There is a lower carbon footprint associated with non-red meat and game (such as wild venison). Buy from farmers' markets or direct from suppliers/butchers with knowledge of farming practices. Look for a regenerative agriculture approach on farms (see page 22) – free-range, organic, grass-fed (not fed on soya and non-waste grains).

Seafood: eat occasionally, but not weekly. Opt for shellfish local to your nearest coastal region (such as clams or mussels) if looking for a nutrient boost. Look for small dayboat catch from farmers' markets, fishmongers or online farm shops, or with the blue MSC ecolabel. Eat lower down on the food chain (pilchards, sardines). Avoid larger farmed or wild fish like cod, salmon and tuna.

Dairy: buy the highest quality that can be afforded, and use sparingly. Look for grass-fed and organic certification. Look out for the Pasture Promise logo, or regional equivalent, that certifies free-range, traditional farming practices instead of indoor, concentrated animal feeding operations (CAFO). Harder cheeses tend to have higher carbon footprints than soft cheeses, as they require more milk. Some studies suggest that goats' or sheep's milk cheeses may have similar, if not higher, associated carbon footprints to cow's milk cheeses. Asking at farmers' markets/farm shops/delivery companies and fromageries will also help in knowing whether animals have been fed on soya and non-waste grains, or as part of a closed-loop farming or regenerative system. Nutritional yeast can be a good vegan cheese flavour alternative (see the Vegan White Sauce recipe on page 139).

Eggs: buy free-range organic, and substitute where possible – certain cakes can easily have at least one egg substituted with a plant-based alternative, like flaxseeds or aquafaba, as featured in the recipes in Chapter 4 (pages 83–112). Look for farmers' markets, farm shops or smaller scale suppliers that can be asked about feed, to reduce likelihood of chicken feed coming from rainforest destructive soya, corn or non-waste grains (even when organic).

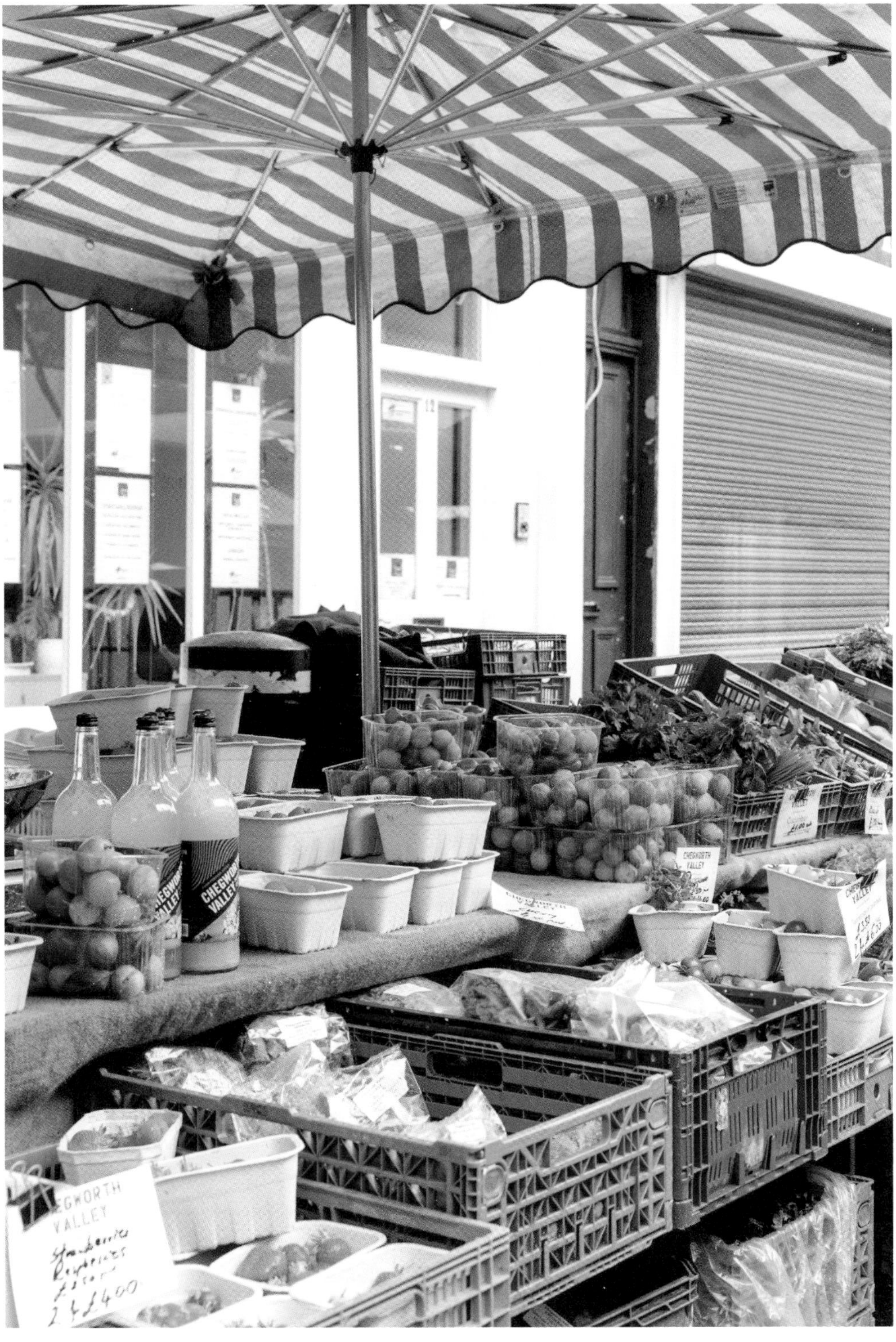
CHEGWORTH VALLEY
CHEGWORTH VALLEY

Oils and cooking fats: diversify purchases – opt for local rapeseed/canola, and organic hemp, grapeseed (grapeseeds are a byproduct of winemaking), argan (as more use will help reforestation of argan forests), coconut and olive oil depending on the dish. Avoid non-sustainable palm oil. Be mindful of highly processed butter alternatives, however, there are simpler options now available, such as Vegan Block.

Nuts: choose agroforestry-grown, organic and local as there are water stress issues associated with growing (with almonds in California, for example). If labelled as originating from developing countries, look for Fairtrade organic. Hierarchy for where to buy: refill/bulk food shops, health-food shops, independent or continental food shops before supermarkets.

Seeds: choose organic and Fairtrade. There is a wide range available and the seeds are highly nutritious, and a great omega oil and protein source, which is especially important when reducing meat and fish.

Sugar: use sparingly. Limit beet sugar due to monoculture and non-organic, soil degradation effects. If using cane sugar, choose Fairtrade, organic and unrefined. Look out for sail cargo, which is increasing in availability. Diversify choices into palm, maple and date sugars, but still look for organic certifications.

Plant-based proteins: there is an increasingly diverse range available. Tofu, made with organic and non-rainforest sourced soybeans is widely available. Quorn pieces and Textured Vegetable Protein (TVP) are a popular meat substitute for specific dishes, but can usually be substituted for pulses – lentil bolognese instead of soya mince bolognese for example. Sausages, burgers, bacons etc. are growing in sales, however, health advice suggests eating ultra-processed foods in moderation, so check the label for additional ingredients. Finding preferred plant-based alternatives as a meat substitution will have a big impact on your personal carbon footprint. Wholefoods, like pulses and beans, are likely to be cheaper, have less packaging and be more carbon-efficient than processed meat substitutes.

Plant-based dairy: choose organic where possible. Oat milk is a popular choice, with some research indicating it is the choice with the lowest associated carbon emissions, and it is often available as a substitute for milk, cream and yogurt. Look for brands that use their leftover oat groats.

Processed food: avoid highly processed foods, instead opting to make your own treats, knowing where the ingredients have come from. This helps to avoid palm oil, and other environmentally destructive additives.

Spices: there is a real difference between the taste and potency of supermarket bought spices, and those who produce spices with quality and ethics. If unsure about origin choose organic. We like Ren's Kitchen in the UK, and Frontier Co-op and Diaspora Co. in the US.

Pet food: look for a good amount of organ meat (10–25 per cent) from different organ sources. Also choose meat that is underutilized (e.g. animals often culled: veal, goat and dairy cow). Another option is insect food, which has considerably lower land use, water use and greenhouse gas (GG) emissions. Many pet chews are underutilized animal parts – dogs love pig's ears, bones and pizzles.

Where is the water and carbon in my food?

Energy is required and carbon emitted throughout the food supply chain. Emissions are often expressed in weight of carbon, or 'the carbon footprint'. Other associated greenhouse gases emitted, such as methane, are associated, these can be converted to a carbon dioxide equivalent (CO2e) to study their impact equally. The water footprint, or 'embodied water', a concept invented by sustainable food systems advocate Prof. Tony Allen, can also be considered.

Production: the production of chemical inputs like fertilizers is carbon intensive. It produces both carbon and other greenhouse gas emissions. A well-known example of this is the production of methane gas in the digestion process of ruminant animals like cows. Less well advertized, is that organic farming can be carbon heavy due to activities such as ploughing.

Processing: the mechanized harvest and processing of food and food products requires energy, water and other chemical inputs.

Packaging: the production (and disposal and recycling) of packaging requires energy, and often directly uses fossil fuels to turn oil into single-use plastic packaging. The incorrect disposal of packaging has other associated negative environmental impacts, such as ocean pollution and plastic leaching.

Transportation: agricultural products are transported from farms to storage and distribution units, then to shops, then homes. This transport is still mostly fossil fuel dependent.

Storage: freezing and refrigeration has substantial associated emissions.

Cooking: cooking, final processing and home storage of food also requires energy and water.

Waste: unsold and wasted food are some of the biggest environmental challenges of the food supply chain. We expend resources on the production, processing, and transportation of food that never reaches our stomachs and packaging that never gets recycled. Even at home, our general, food, and recycling waste often fails to reach its intended destination. Reports from Greenpeace and the Environment Investigation Agency (EIA) show UK plastics shipped to Turkey to be recycled, but dumped and burned instead.

The phrase "reduce, reuse, recycle" is written in that order because when we "reduce", we save resources across the entire supply chain. Our goal should always be to reduce first, then think about how things can be reused, and then recycle or compost if we must.

Logos & farming terms

To get closer to understanding what might be more environmentally sustainable food options, we set out to discover sustainable food stories, and the stories of farmers and growers. It is on the farm that nearly every meal in the kitchen begins. The following guide is intended to help you navigate the certifications and buzzwords around farming practices that produce the food that comes into our kitchens.

Farming logos

Organic

Organic is a food-quality standard that indicates that foods have been grown with limited pesticides, herbicides and fungicides, no artificial fertilizers and specified animal welfare standards. The Soil Association and Organic Farmers and Growers (OF&G) are the certifying bodies in the UK, and adhere to EU standards. In the US, foods must meet United States Department of Agriculture (USDA) organic standards, and other countries have other organic certifiers and symbols. In Europe this is often denoted by the word 'bio', and carries the EU organic logo of a green leaf with star edges.

Rainforest Alliance

This is a non-profit organization working with farmers, forest communities, companies, governments and consumers to protect forests, improve the livelihoods of farmers and forest communities, and promote their human rights.

Fairtrade

Fairtrade International works towards ensuring better prices, working conditions and local sustainability for people who grow and make products. It sets social, economic and environmental standards. Fairtrade covers over 6,000 products from coffee to cotton. In 2018, £166.2 million was paid to producers in Fairtrade Premium and 1.7 million farmers were supported by Fairtrade worldwide. This is an extremely important certification for environmental, economic and social sustainability.

Free-range

Regarding eggs, free-range sets standards for the living conditions of hens. In the UK, free-range means no more than nine hens per square metre with continuous access to daytime runs mainly covered with vegetation. Indoors they must have litter provided, at least 15cm (7 inches) of perch per hen and nest boxes. This ensures they are not kept caged continually. However, it doesn't address diet, beak trimming, male chicks and environmental impacts. Other free-range products to look out for include dairy (Pasture Promise) and pork.

Pasture-raised/fed

Beware – packaging may say 'grass-fed', 'pasture-fed' and 'pasture-raised' but this has no legal meaning or certification. The US equivalent is the American Grassfed Association (AGA), which carries its own logo. Grass-fed doesn't mean pasture-managed; if the animals have been kept on pastures too long, they can do more damage than good. This topic, like so many, is not black and white. We know farmers who only feed their animals on waste products from the farm, which begs the question, is 100 per cent pasture always better? It brings us back to knowing your farmer where possible.

RSPO

The Roundtable on Sustainable Palm Oil label is less common to find, but denotes sustainability efforts for palm-oil sourcing in processed foods. This label can be found on processed foods such as biscuits and some peanut butters. There are many environmental issues linked to non-RSPO palm oil, orangutan habitat loss may be the most well know example.

Slow Food

The Slow Food logo can be found on some food products that are linked to a local membership group. Slow Food started in 1986 in Italy, predating many food sustainability logos and organizations. It was seen as counter-movement to the rise in fast-food culture and this mission has seen it expand globally. Slow Food believes food is tied to many other aspects of life including culture, politics, agriculture and the environment.

Genetic modification (GM)

Genetically modified foods can be defined as organisms (plants or animals) in which the genetic material (DNA) has been altered in a way that does not occur naturally by mating and/or natural recombination. Many pioneering environmental food campaigners had valid concerns about the science of genetic modification, however over time other issues have taken centre stage. Concerns include seed patent laws, corporate interference in plant breeding, global inconspicuous use in growing animal feed and reduced seed sovereignty. Other proponents support the contribution GM seed breeding can make to improved food security.

Certified organic food prohibits the use of GM, and many EU states have also limited their use. There is a GMO free label in the USA. In the UK, foods must say on their label if they contain genetically modified organisms (GMOs) or contain ingredients produced from GMOs.

Farming terms

Permaculture

Often associated just with growing, the principles of permaculture are cemented in design around ecological living. The twelve principles focus on connecting us with the patterns of nature, encouraging us to live low-carbon, yet productive, lifestyles including storing energy and minimizing waste. In food growing, this

encompasses elements such as self reliance (making compost, seed saving); using native ecosystems for water, energy and pest control; and growing a range of produce.

Organic farming

Organic farming focuses on the health of soil, plants and animals, biodiversity and recycling nutrients and resources. Farmers do this by using few, if any, specific, non-synthetic fertilizers, herbicides and pesticides that have been certified as less harmful to the ecosytem, as well as favouring biological control, focusing on high animal welfare (including only using antibiotics when needed), and not using GM ingredients or seeds.

The principles of organic farming are universally the same, however, check your own region for differences in labelling. Organic agriculture has been highlighted in the media for reasons such as large monoculture crops being sprayed with organic chemicals, thus food still being farmed industrially. This is why knowing your farmer and/or supplier is of utmost importance.

Biodynamic farming

Biodynamic farming (also known as closed loop) follows the same principles as organic farming, however, it is more cyclical in nature. This means biodynamic farmers aim to bring in as few external inputs (even organic uses inputs such as manure, straw and animal feed) into their ecosystem, and produce as much as they can themselves. Agreed amendments can be added to the soil, and they often farm in rhythm with the lunar cycle. In many countries it is denoted by a label saying 'demeter', which is the brand for products from biodynamic agriculture.

Regenerative agriculture

Regenerative agriculture is a term that has recently been popularized in relation to growing food and fibres. It does not have any certification or label, but is born from the need to 'regenerate our land' after years of highly mechanical and chemical 'conventional' farming methods. Farmers are now turning to less intensive methods that are focused on rehabilitating our soil. These include, but are not bound by, low tillage, reduction in chemicals, encouraging biodiversity and promoting ground cover. Conventional farms can also be regenerative, if they are working towards moving away from these methods to more rehabilitative practices.

No-till

Tilling or ploughing is the process of turning over the soil, which can be done either by hand or machine. It is often used to kill weeds, and prepares the ground for planting a new crop. However, ploughing releases carbon into the atmosphere – 80 per cent of all the carbon on earth is found in the soil – and also disrupts the soil microbiome and structure. Many farmers, such as Gabe Brown, are now designing systems in which they can sequester carbon in the soil. This usually encompasses no- or minimum-tilling, having animals back on the land and growing different types of produce. There are many farms across the globe practicing no-till, and several farming movements promoting it, such as Groundswell Agriculture in the UK and No-till on the Plains in the US. However, be mindful, no-till (like organic) does not automatically equal sustainable. Farmers sometimes still use chemicals instead of ploughing to kill weeds, although many are trying different methods to eradicate this.

Heritage and heirloom

There is currently no certification for heritage or heirloom produce, therefore anyone can call their produce heritage. There is dispute between farmers, breeders and producers on what heritage and heirloom actually mean in terms of origin and breeding. However, we class them as plant and animal varieties (also sometimes called 'rare' or 'native' breeds) that were/are bred for 'pre-industrialization' farming techniques. There are still great varieties that have been bred more recently, using a combination of old and new

plants and animals for traits such as flavour and low chemical use.

Seasonal

Eating fresh produce when it is in season in the local area will generally have a lower associated carbon footprint than eating it out of season and imported from elsewhere. It is often going to be fresher and tastier, too, because things like fruits and berries have been sun-ripened instead of grown in polytunnels. Seasonal and local often go hand in hand as it needs to be in season local to you. There are many resources to help know when each food is in season, such as the Eat the Seasons website (www.eattheseasons.co.uk), or seasonal food calendars.

Local

Local is an ambiguous term. Most accounts attribute the locavore movement's beginnings to San Francisco in the early 2000s (a person committee to eating food that has been grown in the region where they live), and the US and Canada put it as 160km (100 miles). Food miles refers to the distance a food has been transported before it is consumed. Many environmentalists aim to reduce the food miles of their dietary choices. Some calculations show that local options are not necessarily going to have the lowest associated carbon footprint. This is why local is particularly linked to seasonal, when choosing local as a more sustainable food option. British or Northern European tomatoes out of season may be grown in heated greenhouses, so will likely have a similar carbon footprint to those imported from warmer areas where they are still ripened with the heat of the sun. It is worth noting that when buying animal products, their feed is often imported (soya for example). When considering eating local, this should also extend to what animals are fed. We often like to ask at the farmers' market what animals have been fed on, even if buying eggs and dairy.

Shopping

Carbon in transport

Recent research points to the majority of our carbon footprint associated with food miles coming from individual car trips to the shops, rather than the distance food may travel as overland freight. Reducing trips by car by planning food shopping and buying in bulk where possible, carpooling, taking public transport, cycling and walking are all great solutions for reducing food miles. Getting deliveries or weekly veg box schemes are other ways of dramatically reducing food shopping's carbon footprint. While carbon in transport is an important consideration (both through import/export and our own transport to shops), there are other shopping habits that can be addressed in your move towards a sustainable kitchen.

Where to shop

Where we would like to shop all the time, and where we can realistically shop, are sometimes quite different.

Below is a list of places that we aim to source our food from in order to optimize the environmental sustainability of our ingredients:

- Home-grown organic produce (though it is difficult and unrealistic for most to live entirely on home-grown produce).
- Farmers' markets and farm shops.
- Health food, wholefood and refill shops.
- Independent, continental and corner shops (now supplying a lot of organic and diverse vegan options).
- Online farm produce and organic veg box schemes.
- Online sustainability-focused wholesale suppliers.
- Local markets (fresh and unpackaged, but not usually organic, exclusively in season or local produce).

See page 36 for a list of places that we source kitchenwares from.

These are places that we try to avoid using for the bulk of our shop:
- Large chain supermarkets.
- Non-environmentally focused brands.

Food waste reduction apps and resources

There are a growing number of apps and platforms designed to help redistribute food that may otherwise be wasted. Olio and Too Good To Go are two of the most popular ones in the UK (Food For All in the US and ResQ in Europe), and they have free and heavily reduced food posted everyday. This comes from restaurants, bakeries, food shops and individuals who've baked too much or are doing a cupboard clear out. The apps enable citizens to either pick up surplus food for free, or purchase it at a major discount. They also help save money, which helps to justify spending more on other organic or more expensive sustainability-focused produce. Organizations like Feedback and This is Rubbish in the UK, and City Harvest and Oz Harvest in the US and Australia, respectively, are a great starting point if you want to learn more about the global food waste issue, and the systemic changes needed to deal with the root causes of food waste and loss.

Foraging

Foraging, or picking wild foods, is a liberating and empowering act. If you've been brought up to pick wild blackberries or apples as you walk along country roads, you may have always been a forager and not even realized. Of course, plant identification can get very complicated, and can be dangerous, especially with mushrooms. For this reason, this is not a guide to foraging. This is our homage to foraging as an engagement tool for embracing more sustainable food systems and bringing nature into your kitchen. When we embark on foraging walks in and around London, it helps us to connect with the natural world. Being able to identify edible plants helps us to understand seasonality as well as appreciate the bounty and biodiversity of where we – even in a megacity.

Here are our top tips for foraging:
- Never eat any wild food without first being 100 per cent sure of its identifty. Start simple; blackberries, apples and pears.
- Research the edible plants in your area; find out if any are endangered, how to pick them and be wary of spreading invasive species. Your best bet is to find a reputable foraging class or walk.
- Get a good foraging guide and familiarise yourself with differnt leaf, stem seed and fruit shapes and structures.
- Take no more than you can process or consume immediately, before it goes bad. Many places in the UK, such as some Woodland Trust land, allow foraging only for personal use, but not commercial purposes.
- Take care not to trample biodiverse areas and avoid uprooting any plants.
- Seek permission from landowners, or forage on your own load or public land for which you know you can look up the foraging rules.
- Only collect leaves, flowers, fruits and seeds where they are in abundance, and leave a good proportion in place. This will allow them to grow back and drop seeds to make more plants, which provide food for birds and other wildlife. We only collect a few types of mushrooms that we know in our area, and only pick if they have opened their caps, as they are more likely to have dropped their spores.

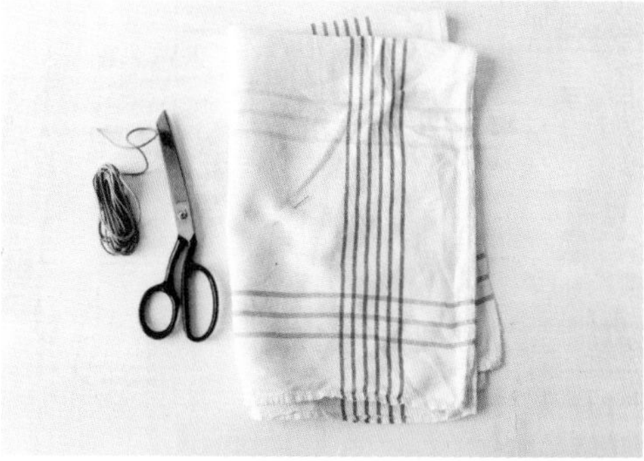

1

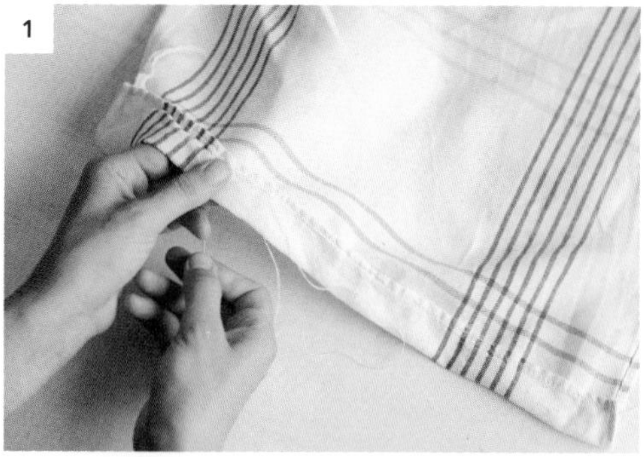

2

3

4

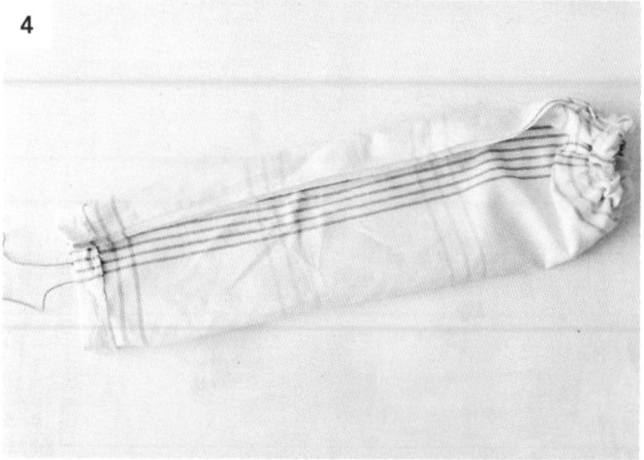

5

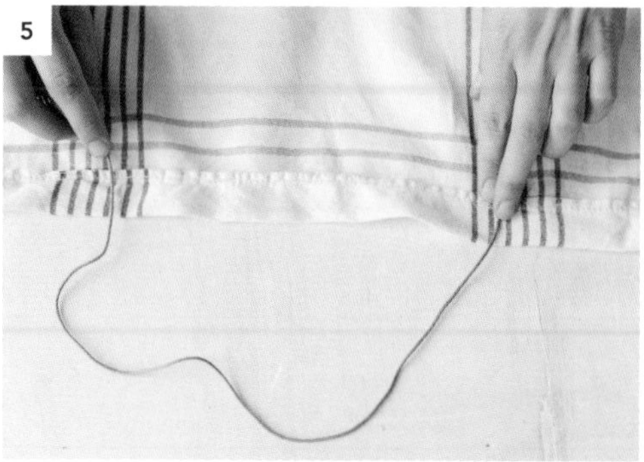

6

7

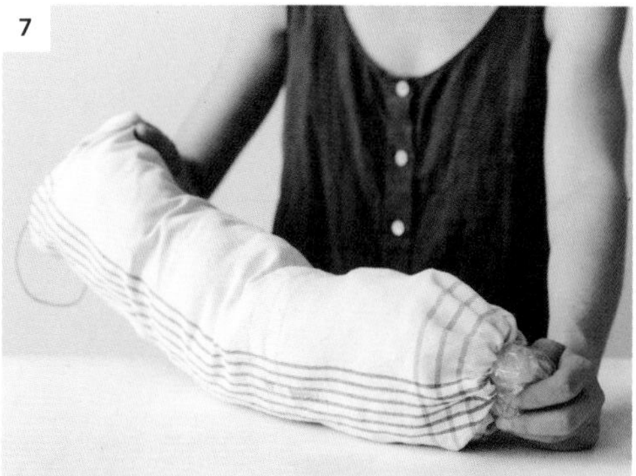

Make Your Own: The Bag Bag

It is now the norm in many countries that single-use plastic shopping bags have to be paid for, as well as more durable 'bags for life'. Fair enough, when you consider that an estimated one trillion single-use plastic bags are used each year worldwide, many of which end up in oceans and tangled in trees. In an attempt to reduce our inconspicuous collecting of plastic bags, and to help us remember to bring a reusable bag out with us, we have a bag bag.

I'm sure many of us have ended up with more cotton tote bags than we can get round to using. By having a bag of clean, reusable bags hung on the back of a door, in a hallway or on the coat rack by the front door, it makes it hard to forget to pick one up when leaving home. Regardless of whether we plan to go to the shops or not, we always seem to end up with something to bring home.

This is a guide for making your own niftily salami-shaped elasticated bag, like Sadhbh's mum made. Alternatively, just fill a bag with bags and take them out of the top. Ta da! Easy peasy.

Materials

- Roughly 45 x 56 cm (18 x 22 inches) piece of durable thick fabric, or even a teatowel.
- 1 x 20cm (8 inch) piece of 5mm (¼ inch) wide elastic
- 1 x 20cm (8 inch) piece of twill or ribbon
- Safety pin, pins, thread, iron, sewing machine (or hand sew)

Method

Sew a hem along the shorter edges of the rectangle of fabric [1].

String the elastic through the bottom hem by popping a safety pin through the end of the elastic, and sliding it round until it meets on the other side [2]. Tie the two ends of elastic together [3].

Sew one end of the ribbon a quarter of the way across the top hem, and sew the other end three-quarters of the way along [4]. The lengthwise inner seam can then be sewn together to create a cylinder [5].

Tie together the ends of ribbon, and tighten the two ends of the bag [6]. The tighter you pull them the narrower the neck and base of your bag bag get. This is where you will put bags in at the top, and take bags out at the bottom [7].

Interlude: a tea break

The fundamentals of *fika* are really important to us at TSFS. A Swedish tradition, *fika* is difficult to directly translate, but essentially it means to have a break, drink coffee, eat cake and have a chat. This is why we love it so much. We work hard and play hard; you have to with the fast-paced events we do. However, this is more than just a cuppa, it symbolizes slowing down and taking your time – which is important in the world of activism, as otherwise you feel you can never stop!

Of course, for us tea breaks often bring up our favourite question ... how can they be as sustainable as possible? Here's our how-to guide.

Tea choices

Origin

When we talk about tea from the *Camellia sinensis* plant (that includes black tea, white tea, yellow tea and green tea) there are several different certifications (see pages 20–23) that we look for. According to Ethical Consumer the majority of our tea in the UK comes from smallholdings in Kenya, but some comes from plantations in India. The conditions on the plantations are demanding, with lower than average wages, and certifications do not seem to be helping, whereas certifications have shown positive impact to the environment and people in Kenya. Therefore, we choose Kenyan tea where possible.

Unless you happen to live in a coffee or tea-producing region, it may be tricky to know the farmer, so do research online. The internet has given us the power to know more about the origins of our food, and the more we demand transparency and sustainable practices the more companies will have to abide by them.

Teabags

Make sure your teabags are biodegradable or compostable, with no plastic, and that any tags are paper. These can then be put in food waste or composted.

Packaging

Make sure any outer packaging is recyclable.

Filling your kettle

Only fill your kettle with the amount of water you need. The easiest way is to fill the cups you are using with water and pour them in. This saves energy and water.

Milk

See page 16 for a guide to dairy and milks.

Making your own herbal teas

Herbal teas can be made from fresh or dried leaves and flowers – some leaves such as parsley and basil are best fresh. If you are familiar with wild plants, foragables like nettle, blackberry and raspberry leaf, and yarrow are popular wild teas (see Foraging, page 26). Others lend themselves well to being dried, such as mint, lemon verbena, lemon balm and lavender.

To dry, hang your leaves/buds/flowers upside down in a warm place. You can cover with a paper bag if need be. The teas are ready when they feel dry and crumble in your hand. Store them in an airtight container.

Coffee

Origins

When choosing coffee, Fairtrade and organic are our go-to certifications. We also look for shade-grown coffee, which is grown under a larger canopy of trees, protecting the land against deforestation, biodiversity loss and soil erosion. Choose certifications such as Rainforest Alliance to support this.

Decaf

There are four mains ways to decaffeinate coffee: indirect-solvent, direct-solvent, Swiss Water and supercritical CO2. We opt for Swiss Water method due to the non-chemical process. It is also the only method that can be used for organic coffee.

Packaging

We look for minimal packaging, which can be recycled. Be mindful that any bags are not lined with single-use plastic. If so, they can go in an ecobrick. Coffee pods or capsules are unnecessary packaging, even if recyclable you can avoid this by buying your own loose coffee.

Sweet treats

See pages 16–18 for our thoughts on flour, sugar and fats. Although fika buns are often laden with sugar, butter and white flour, there are so many options!

Alcohol

Booze is, of course, arguably not an essential part of the kitchen. But many would insist otherwise. We do use alcohol in our cooking, and like a tipple with our meals, and we serve forage-infused cocktails at our supper clubs.

We've sometimes struggled to find sustainability-focused alcohol brands, but their market share is gradually increasing. Our original matchmaker, Tristram, is also the founder of Toast Ale, an innovative London-based beer made from surplus bread. It's recipe is open source so anyone can try it for themselves.

We prioritize buying organic and Fairtrade wines where possible, or biodynamic wine, which has a lot of overlap with organic. Natural wine is also organic and sometimes biodynamic. A couple of our favourites are Tinte natural wine and Davenport organic wine.

We've also purchased from The Sustainable Spirit Co, who supply spirits in a refillable 'eco-pouch'. This reduces packaging by 95 per cent.

Homebrew beer from kits or from scratch is a truly admirable endeavour and definitely has the lowest food miles. Sadhbh's dad and uncle are master homebrewers, creating such concoctions as chocolate stout, oak leaf wine and parsley wine (using the herb garden glut), foraged elderflower champagnes and elderberry wines and beers. Check out our pal Jo of Edible Flower Brews for more inspiration and recipes.

Creating a sustainable kitchen

We like to deep dive into sustainability considerations for all purchases and products, including equipment, furnishings and aides in our kitchens – not just food. As desirable as it used to seem to us to have a kitchen with all the latest gadgets, matching pristine utensils and bright and shiny new everything, we've challenged ourselves to resist buying new where possible and make do with what we've got, borrow or find it secondhand. We question whether something is really a need or a want, to help guide our purchases. We noticed that after our student years, as had a little more disposable income in our graduate lives, it was easy to fall into the consumerism vortex, buying more and more with less thought about each purchase. To help us switch our mindset and move away from that short-lived high we associated with buying something new, we've adapted to see it as a triumph when we don't buy something, and now feel proud of ourselves when we find an alternative to buying a new kitchen item. Just as we build other healthy habits, like doing exercise. Many people can admit they once found exercising boring, or a lot of effort. But we know it's healthy, so we learn to love it, and eventually look forward to exercising, whether that's attending a fitness class or hiking a mountain.

In our hope of helping readers along in making more informed decisions about what comes into their kitchens, this next section covers sustainability issues of the kitchen kit we use, and think you might, too.

Gathering kitchenwares

We have to admit, buying new things is not something we do often. Whenever you buy something new it always has a carbon footprint, no matter how sustainable the business. We pride ourselves on building our kitchens out of as much secondhand equipment and furnishings as we can. Acquiring secondhand equipment just means you give it a good clean and sterilize it before using. This goes for furnishings and white goods too – tables, chairs, fridges, washing machines – there is very little you can't find secondhand.

We focus on the following options when sourcing kitchenwares:

- Secondhand from Ebay, Gumtree, Freecycle, Facebook Marketplace (make sure the seller ensures electrical items are in good working order, and can be returned if not. Still boxed can be the safest bet).
- Charity and secondhand shops. Free collection of donations can also be booked with many charity shops.
- Hand-me-downs from family and friends. People are often happy to offload the likes of a bread machine or ice-cream maker they bought and used a few times. A 'wanted' post on social media might just find you what you're looking for at half the price or even free.
- Car-boot sales, yard sales/giveaways, flea markets, vintage and antiques markets.
- Artisans, craftspersons and bespoke makers often source with sustainability in mind, and can ensure traceability of materials used.
- Eco-conscious brands that use recycled, refurbished or upcycled materials, have replaceable parts and have sustainability commitments published on their website.

However, if you do need something that you can't find secondhand anywhere, here is a little guide to our choices:

- Buy the top quality you can afford from a well-known brand, as it's more likely to come with a guarantee and to last longer.
- Buy from the Ethical Superstore or a website with environmental rating guidance. This may cover BPA-free plastic, FSC wood and energy efficiency ratings.
- Buy A+++ rated whites good (EU) or Energy Star certified (USA, Canada, Japan, Switzerland, China) where possible.

Ergonomics: design waste out of the kitchen

Ergonomics is the fancy word for the study of efficiency in a working environment. In home economics class there was a section in our textbook about 'the ergonomic triangle', aka 'the kitchen triangle'. Having the cooker, sink and fridge at three points of an imaginary triangle in the kitchen can really help with efficiency. Increased efficiency should mean decreased time and food waste, and increased willingness to cook healthy food from scratch.

Years of working in commercial and home kitchens has highlighted to us just how much designing things to be easy means we are more likely to do them. For example, if the sink is a swivel away from the hob, it makes it easier to quickly rinse out a pan and reuse it, and leaves us with fewer dishes to wash at the end of the meal. If a fridge is in another room, or hard to get to, it's more likely that putting leftovers away at the end of the day or meal prep session will be forgotten, so spoilage and food waste might be more common.

There are some very easy storage decisions you can make, such as keeping your kitchen knives and chopping boards near your work surface, your cutlery and crockery near where you wash up and your mugs near the kettle. These little choices will enable you to move easily around your kitchen.

Setting up your bin systems

The same theory applies to the bin system in your kitchen. Having easy-to-open, accessible bins for general waste, recycling and compost make it easier to put waste in the right place. This might be especially true if living in a shared house with flatmates or family who aren't quite as green minded as you. No one wants to be the eco-nag, but if you can set up a system that makes it as easy to recycle or compost as it is to chuck it all in general waste then you might find people pick up the habit more easily.

Here are some ideas for where to position your waste organization:

- In the cupboard under or beside the main preparation counter, so food waste can be scooped straight in.
- Lined up next to the main prep counter, preferably with all three bins together (general waste, recycling and compost). There are bins that have compartments of varying size, suitable for different types of waste.
- Having a countertop food waste caddy. Your local council may even provide you with one, and biodegradable food waste and recycling bags might be available for free from a local library or council offices. Check your council website or ask your landlord.

Creating a kitchen you want to cook in

One of the most important, often overlooked, aspects of creating a sustainable kitchen is building one that you want to be in. If you don't want to cook, eat and chat in there then all the hard work in creating the space is moot.

We chatted to designer Freya Rose Tanner who said it's about making conscious choices about what enters your space. 'Take a look at what is serving you, is that dustpan and brush you've replaced three times in the last year serving you or the environment?' The best design is when you don't even notice it, you can move around the kitchen with ease, the colours make you feel at home, and Freya advises the objects should meet you 'physically, energetically, emotionally'.

This does mean any purchases should be made with intent. 'Enjoy your abundance, know your excess' she remarks. As many objects can speak to us, we just have to think mindfully about what we actually need, which sometimes trumps what we want.

Equipment

When we do have to buy something new, we have criteria we look out for. We like wood, however, would look for some assurance of environmentally conscious foresting practices. The Forestry Stewardship Council (FSC), despite some limitations, is a good place to start. Bamboo is a fast-growing plant (it's actually a grass), and many eco kitchenware brands supply bamboo utensils. Bamboo has many merits, but it's important to remember it can still be grown as a monoculture cash crop, and can still displace native forests and wildlife.

Utensils:

- Buy for longevity. It's often better to buy that tin opener from the homeware shop rather than the pound shop or supermarket own brand. It may cost a bit more, but if it's a well-known brand, it might be worth the investment as it will likely last longer.
- See if you know anyone with doubles, looking to re-home stuff. We've often acquired a wooden spoon and spatula when someone's doing a clear out.
- Go for the mismatched kitchen utensil look – not being fussy about a matching look will help when sourcing quality secondhand items.
- If buying new, choose wood or metal over plastic.
- Stainless steel is much more durable than aluminium, so is generally worth the price difference for utensils.
- Silicone can be a good alternative to plastic. Silicone spatulas, pastry brushes and spoons are easy to clean and pretty durable.

Chopping boards

For home use, we prefer wood to plastic. Wood is more durable, aesthetically pleasing, easy to clean and there is some research to suggest it has antibacterial properties, so is a hygienic choice.

Look for secondhand wooden chopping boards. They'll be a lot cheaper than a nice new, heavy chopping board, and can always be sanded and oiled with linseed oil or even cooking oil. Alternatively, look for handmade boards by local craftspeople who can guarantee the provenance of their wood, or those from FSC-certified or sustainably managed forests. The illegal timber trade for the likes of kitchenwares and furniture is a huge threat to biodiversity conservation and the fight against climate change.

Crockery and textiles

As with kitchen appliances and cookware, there are some simple steps to consider when acquiring tableware:

Choosing crockery

- Look out for pre-used crockery. Car-boot sales are notoriously good places to find a diamond in the rough, or crystalware in a car boot.
- Buying good quality that will last longer is key to reducing the need to replace crockery and glassware. Heavy tumblers, plates and bowls with sturdy edges will be worth it.
- Look out for glasses and porcelain made with recycled materials. Terrazzo crockery seems to be gaining popularity with sustainability-focused kitchen brands. This composite material of marble or granite set in polished concrete is more well known as a tile, floor or kitchen counter surface. Some terrazzo contains up to 80 per cent recycled materials from stone and marble processing industries, and some come LEED accredited (Leadership in Energy and Environmental Design).
- Making your own pottery at pottery lessons is a fun and impressive way to acquire your dream, handcrafted pottery set.
- If making from scratch isn't your thing, you can always upcycle some ugly mugs with some dishwasher-safe ceramic paint and a little creativity. You can make your mismatched collection feel a bit more uniform.
- Crockery can be hired for bigger events, and usually at a lower cost than buying lots of cheap crockery that might only be needed once. Glasses for parties can even be hired for free from some supermarkets or drinks merchants. You only pay for glasses that break.

Choosing fabrics

Dishcloths, tea towels, bags and table linens are just some of the fabrics that feature in the kitchen. We sat down with good friend Hetty Adams, founder of HENRI and The Modern Sewer, who talked us through her sustainable natural fabric choices. She is an advocate for the link between farming, food and fashion, as we often forget that most natural fabrics start life as seeds.

For information on cleaning cloths, pads and scourers see pages 153–65.

Cotton: cotton is one of the thirstiest crops – when conventionally grown it is highly unsustainable largely due to water usage, and the vast amount of chemical pesticides and fertilizers used. Henri uses organic cotton grown using rain water, which can reduce the water consumption by up to 90 per cent and reduce the chemical burden. However, check with your supplier about traceability as this is not always the case. If buying cotton for the kitchen, we choose organic.

Linen: made from the fibres of the flax plant, linen is strong and absorbs moisture. Flax plants grow easily and many parts of the plant can be used. It can withstand high temperatures, making it great for tablecloths and tea towels.

Hemp: hemp is one of the most sustainable fibres. Grown from the cannabis plant, hemp can grow almost anywhere and is a virile plant. The fibres were once known as a very rough fibre, however, modern production techniques have made it available in much lighter weights.

Nettle: this is another easy-to-grow plant. Used the same way as linen, it provides long, coarse fibres used for weaving fabric.

Soy: soy fabric is produced from food waste, being made from the hulls of the soybean. It produces a soft, lightweight and absorbent fabric.

Bamboo and lyocell: bamboo is the world's fastest-growing woody plant (it is not classed as a tree), a renewable resource and can be spun into fibres, so seems like the ideal material for sustainability. However, due to the toxic solvent used to break down the material, lyocell has been deemed a better choice. Lyocell is made from fast-growing, hardy eucalyptus trees and uses a non-toxic solvent. It can also be FSC certified.

Recycled materials: we are strong advocates of fabrics made from recycled materials, and of recycling your own materials to use in the kitchen. Old shirts or torn clothing can be used for cloths or sewn into tea towels.

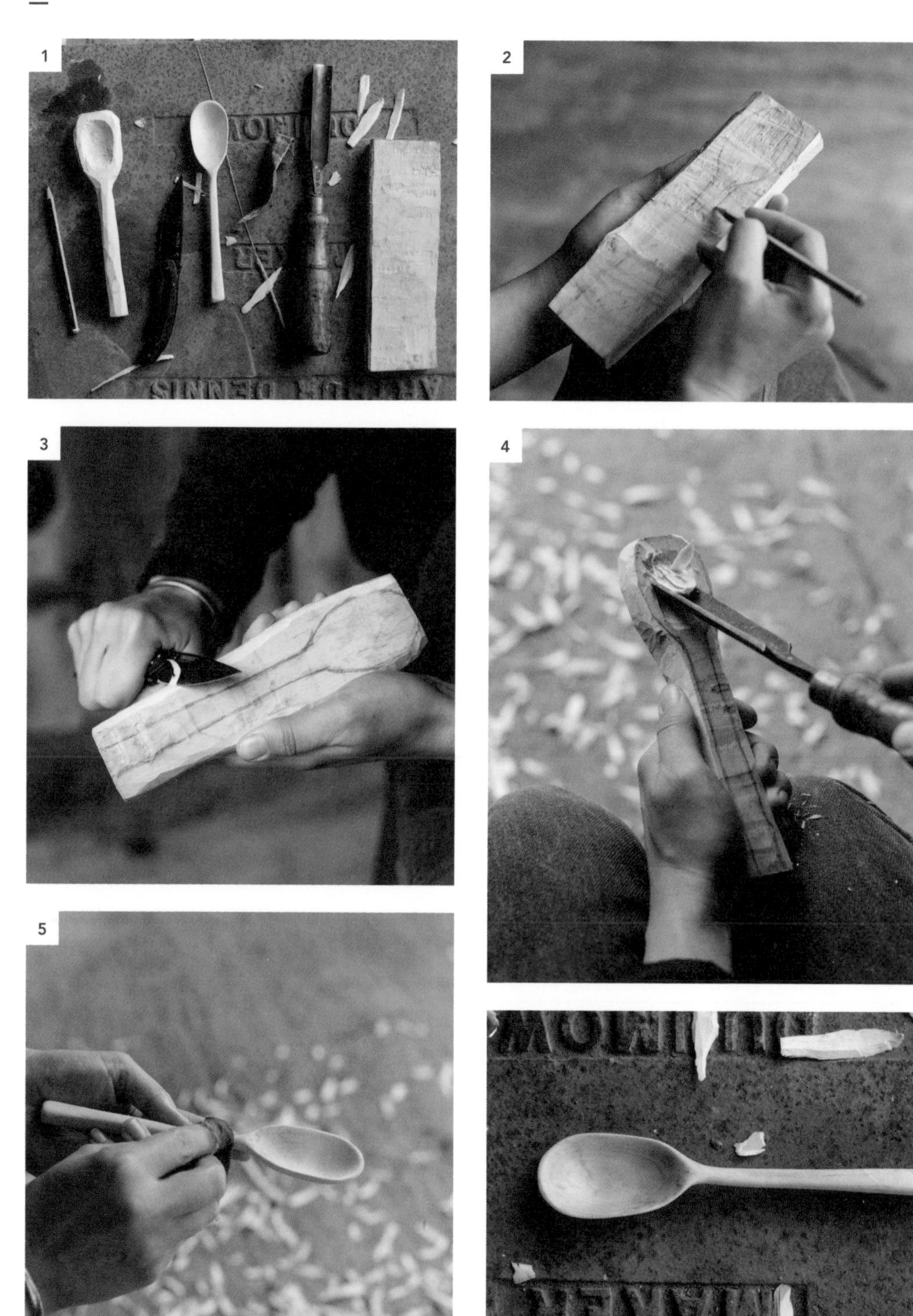

Make Your Own: Whittled Porridge Spoon

This might not be the quickest way to furnish a kitchen, but is it satisfying, low impact and impressive!

Materials

15 x 5 cm (7 x 2 inch) piece of straight green wood (a fresh branch from a tree, or one that has been cut recently so has not had time to dry out. Alternatively, you can buy a 'blank'.
Sharp carving knife
Hook knife (or a rounded gouge), (optional)
Pencil or marker
80, 120, 220 and 400 grit sandpaper
Linseed oil (or any cooking oil)

Method

Firmly holding the piece of wood in one hand or a vice, put your thumb against the back of your carving knife and place the blade 1 cm (½ inch) or so down the branch, against the bark and facing away from you.

Push firmly with your thumb, chipping off the bark and a chip of wood just below it. Repeat, rotating the branch, until you have removed all of the bark.

With a pencil or marker, draw the shape of the spoon you want to carve onto the wood – this silhouette will work as a guide for the rest of your carving. It's best to start with a big and chunky spoon. Refinement comes after some practice **[2]**.

Now repeat the chipping movement, whittling little chips of wood from the blank you've produced, and aiming to carve out the shape of the spoon you've drawn on. This may take many hours and may make your thumb feel a bit sore, or you may have strong hands and find it quite easy. **[3]**

When it comes to carving out the bowl of the spoon, you may want to switch to a hook knife or rounded gouge. It's important not to dig too deep, as you'll carve from the back too, and meet in the middle. You need to keep the wood thick enough so as not to end up with a weak, too thin spoon **[4]**.

When you're finished and happy with the shape of your spoon, however crude, you need to start sanding. Begin with the 80-grit sandpaper and work your way up to the finest grit to get the wood nice and smooth.

Rub with oil and leave to dry out slowly, not too close to a heat source as it may crack **[5]**.

Energy & water use in the kitchen

Energy and water demand are entwined in almost every activity in modern society. Both have big ecological footprints and are factors we consider when buying and sourcing kitchenwares and foods, too. In an effort to not get too bogged down with the complexities of these big topics, we've compiled some energy and water saving tips for working in the kitchen.

Tips for reducing energy use in the kitchen:

- Fix dripping taps. A dripping tap can waste 5,500 litres (5,800 quarts) water per year, which is enough to fill a paddling pool every week for the summer.
- Use a smart meter.
- An electric water heater may be more energy-efficient than a stored water heater that heats up at certain times of day and goes off to a thermostat, regardless of whether hot water is needed then or not.
- Run the eco setting on the dishwasher. Wipe plates first with a damp cloth instead of rinsing under running water. The old 'mopping up sauce with a crust of bread' also helps make plates pretty clean and dishwasher ready.
- If you live in a house or flatshare, cook meals together, to share the energy cost as well as pots used and needing washing.
- Don't leave the oven door open for too long, to keep the heat in. Same goes for the fridge and freezer, but to keep the cold in.
- Try not to have the fridge right beside the cooker or oven as it has to work harder to stay cool, and uses more energy.
- Using the kettle tends to use less energy than boiling water on the stove/hob, and try to only boil as much as you need.
- Defrost your fridge and freezer regularly to keep it working efficiently.
- Switch to LED or energy-efficient lightbulbs. But remember to switch off lights and appliances when not in use.
- Switch to a green energy supplier. There is plenty of guidance available online about how to do this, and plenty of free comparison websites and services to shop around.
- Be mindful of energy ratings when sourcing appliances. Choosing high energy ratings for appliances that are used a lot (cookers) or continually running (fridges) is important for not only minimizing energy usage, but also saving money. Thinking about the size of the appliance and the setting is important, too. Choose the highest-rated appliance you can afford, whether buying new or used.

Planning ahead

Storing produce correctly is a sure-fire way to extend the life of your food and minimize waste. The overarching tips for storage are:

- Be mindful not to overfill your cupboards and storage spaces – that's when items get hidden and go off.
- Keep your fridge (1–4°C/34–39°F) and freezer (<-18°C/-0°F) at the right temperature.
- For best energy efficiency, don't overfill your fridge, keep away from heaters and repair any broken seals.
- Avoid storing anything in direct sunlight.

Storing leftovers, different kitchen rolls and wraps

Cling film or plastic wrap: we try not to use single-use plastic wherever possible. For covering foods or wrapping up sandwiches for trips we use washable beeswax wraps (see page 51), reused plastic bags or tupperware tubs and boxes.

Greaseproof or baking paper: baking paper gets reused for as long as it will hold up for, and can be used repeatedly for cookies and breads that leave it pretty clean. Used oily greaseproof paper makes a great firelighter for a barbecue or campfire.

Tin foil or aluminium foil: Foil mostly gets used for cooking meat, so by eating less meat we find we use less tin foil. We cover veg in oil before roasting, so never line our trays with foil. If we've used foil to cover a tart that's cooking too quickly on top, if it doesn't get food on it, then we carefully remove it, store it and reuse it multiple times. As foil is made from the non-renewable mined metal aluminium, if clean and dry, it can be balled up and recycled, just as drinks cans are.

Tupperware or plastic tubs: These tubs are designed to be used again and again, even tubs from takeaways. They have to be food safe, and can usually be microwaved, so save on dishes if you're ever heating up leftovers for a quick meal. We also sometimes store leftovers in glass jars in the fridge, or in a bowl with a plate over it. Storing leftovers and remembering to eat them is one of the biggest ways we can reduce household food waste.

HOW TO STORE PRODUCE

Here we break it down a little further:

Berries, plums, grapes: store in the fridge.

Apples: store out of the fridge away from other sources of ethylene (high in ripening bananas), which catalyses ripening.

Herbs: store tender herbs (e.g. parsley) wrapped in damp paper towels, or you can trim the ends of bunches and store in water in the fridge. Hardier herbs (e.g. rosemary) can be stored in paper bags.

Roots and tubers (e.g. beetroot, potatoes): store in a cool, dry place in paper bags or boxes, or in the fridge crisper drawer wrapped in paper. Leave unwashed.

Onions, garlic and winter squash: store in a cool, dry place in paper bags or boxes. Leave unwashed.

Peppers and mushrooms: store in a paper bag in, in the fridge, unwashed.

Broccoli and other leafy greens: store wrapped in moist paper towels in the fridge.

Carrots, parsnips, citrus: store in the fridge, but they but will last in any cool, dry place.

Celery, cabbage, courgettes: store in the fridge, wrapped in paper in the crisper draw.

Cucumber, nectarines, peaches, apricots: keep at room temperature – they dislike the cold!

Bananas, kiwi, mango, pears: can be left out of the fridge until ripe and then moved in to inhibit further ripening. When cut they should be stored in the fridge.

Bread: store in a paper bag in a cool, dry place.

Eggs: store in cool, dry place, or if you have room they can go in the fridge. You can check your eggs by placing the egg in a cup of water, if it sinks it's still good. If it floats, maybe ready for compost.

Dairy: store in the fridge.

Non-dairy alternatives: if bought at ambient, then store at ambient until opened, then refrigerate. Otherwise store in the fridge.

Nut butters: can be stored at room temperature.

Leftovers: always store leftovers covered in the fridge.

Rice, pasta, dried beans, tins: store in a cool, dry place.

Oils: store in a cool, dry place away from direct sunlight.

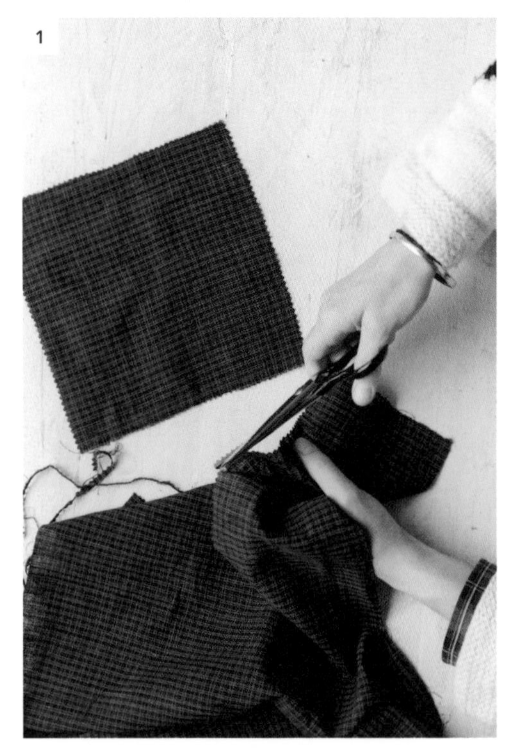
1

2

3

4

Make Your Own: Beeswax Wraps

Fabric wraps coated in beeswax have become the poster child of home plastic-reduction efforts, and no wonder. They're easy to DIY using upcycled fabric, make great gifts and are restorable when the wax starts to wear off. You can even use the ends of pure beeswax candles to make these simple beeswax wraps.

Materials

- Baking tray
- Greaseproof paper
- Non-stretchy, 100 per cent cotton fabric (such as a former curtain, thick pillowcase or tea towel)
- Scissors or pinking shears
- Beeswax pellets, sheets of beeswax, grated or chopped chunks of a block of beeswax or unused ends of pure beeswax candles

Method

Preheat the oven to 180°C (160°C fan/350°F/Gas 4). Line a clean baking tray with greaseproof paper.

Cut the fabric to your desired size – a 20 cm (8 inch) square works well **[1]**.

Place the cotton fabric square on the greaseproof paper. Spread a handful of beeswax pellets or grated beeswax evenly over the fabric **[2]**.

Transfer to the hot oven for 2–3 minutes until the beeswax has melted into the fabric **[3]**.

Remove it from the oven and leave it to cool before peeling the fabric off the paper. Alternatively, lift the fabric up carefully when still warm, allowing the excess wax to drip off, and then hang the fabric up using clothes pegs and leave to dry.

Repeat with different dimensions of fabric to achieve different size wax wraps to make a collection.

VARIATION: you can melt wax in an old saucepan and use an old brush to apply wax to the fabric rather than melting in the oven, but this is messier to tidy up after.

1

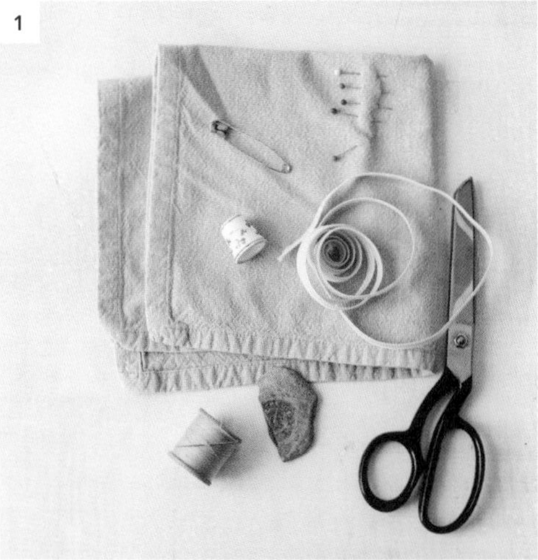

2

3

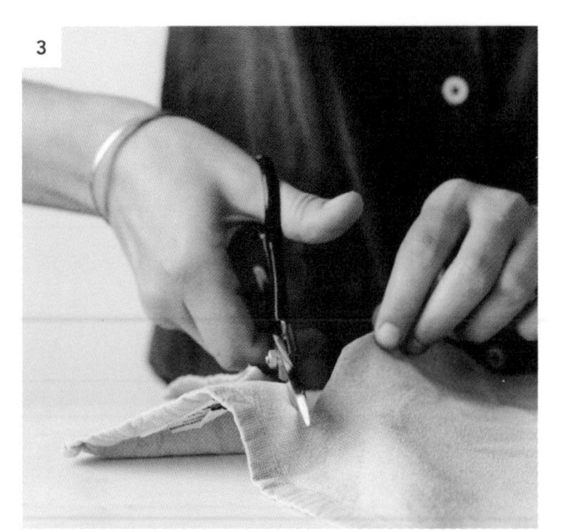

4

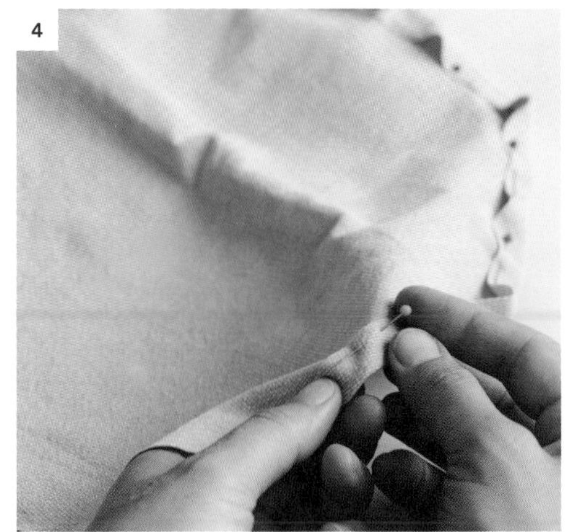

5

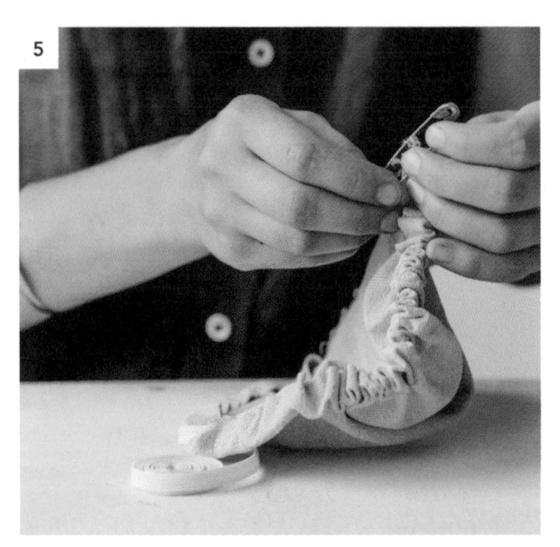

6

Make Your Own: Bowl Covers

Another cover story. Many bread bakers will have learnt to cover bread dough with a clean, damp tea towel, to avoid it going crusty as it rises at room temperature. Making a simple fitted bowl cover means you'll never have to use a clean tea towel, or a stained or torn old one and wonder if it really is clean enough to be in contact with food.

You could use a beeswax cover (see page 51), but having a plain fabric one means it can be kept moist and you can shove it in the washing machine. A fitted cover avoids having a trailing damp tea towel hanging off your bowl and the risk of the cloth falling into the bread.

Materials

Bowl to measure around.
Fabric for the cover
Length of 1 cm (½ inch) thick elastic band, measuring at least the circumference of the bowl
Safety pins and pins
Needle and thread

Method

Start by drawing around the bowl you want to cover. Draw another circle around the outline that is 4 cm (1½ inch) wider **[2]**.

Cut around the wider outline **[3]**.

Pin or tack a 2 cm (¾ inch) hem all the way around the edge of the circle of fabric, leaving a gap at one side that is large enough to fit your elastic through the hem. This will pucker, because you're folding a larger circumference onto a smaller circumference, but that's OK. Just try to spread the puckers, or pleats, out evenly **[4]**. Sew the hem in place.

Hook the end of the length of elastic with a safety pin **[5]**. Use the closed safety pin to thread the elastic through the hem and tie or sew the two ends of the elastic where they meet, so that it pulls the hem together and starts to resemble a shower cap.

Place over the bowl to check it fits snuggly and you're all set **[6]**!

TIP:
You could use elastic from decommissioned pyjama bottoms – we try to save ours, balled up in the sewing kit, ready to upcycle for projects like this.

Reusing jars & tubs

This is knowledge that two generations ago, anyone who ran a home kitchen (usually the woman of the house) would have known as second nature. Preserving was part of feeding yourself and your family a wider range of foods through the year.

Saving jars from bought or gifted preserves is a great way of acquiring bottling materials, and one of the best opportunities to 'reuse' when looking to limit what gets binned and goes for recycling from our kitchens.

Tupperware with clip-lock sides can be great for making pickles and cold ferments, too. You shouldn't put anything hot into a plastic container in case it degrades the plastic, but we often make a quick pickle or ferment a kraut in a food-grade reusable container or tupperware. It just needs to be sterilized by soaking in boiling water for a few minutes before using.

Guide to sterilizing jars and hot-bottling preserves

- Wash jars and lids and rinse soap suds out in hot soapy water. Place them on a baking or roasting tray and heat in an oven preheated to 160°C (140°C fan/325°F/Gas 3) for 10–15 minutes.
- We use sterilized tongs (silicone tipped worked well for this) to lift the jars and lids out of the oven.
- Put hot preserves into jars when they're hot out of the oven.
- Sterilize a heatproof jug, ladle and funnel if using, to move the preserve from the pan to hot jars. A funnel or ladle with a lip helps to prevent any preserve spilling over onto rims of jars, especially for chunky preserves.
- Fill the jars almost to the top but leave ½ cm (¼ inch) gap as the lid needs to fit on without squashing any out. If lids of jars don't look as clean as you'd like, you can use a greaseproof paper disc to create a sterile seal between the jar and the lid.
- Lids should be placed on while everything is still hot. Be careful not to touch the inside of the lids. We use tongs to lift up lids or hold jars if they are too hot, while we screw on the lid.
- Preserves hot-bottled in sterilized jars should keep for at least six months, and should be stored in a cool, dark place. Check the recipe you're following, if following one, as it may have more specific shelf-life instructions.

1

2

3

4

Make Your Own: Ecobrick project

AJ, head chef at The Castle Climbing Centre – one of the most sustainable kitchens we've ever worked in – first introduced us to this nifty way to think about our plastic use. Making an ecobrick from your plastic waste allows you to see it compacted and think about building projects that could utilize our non-recyclable kitchen waste. We've looked up some innovative building projects all around the world that use these. Ecobricks are being utilized by communities in certain less economically developed countries where there may be an abundance of single-use plastic waste and plastic bottles, but little infrastructure to dispose of it in landfill. There's even a Global Ecobrick Alliance.

The household waste that you use for the ecobrick should primarily be that which cannot be recycled and is malleable and scrunches up small enough, but you can make an ecobrick from anything non-biodegradable and dry.

Materials

Selection of clean and dry household waste
Plastic bottle
Stick or wooden spoon

Method

Twist and scrunch the household waste and insert it into a plastic bottle [2]. Compress it as tightly as you can by poking it down with a stick or the handle of a long wooden spoon [3].

Repeat the first step with all non-recyclable and single-use plastic waste – make sure your bottle is unsquashable as it becomes filled with compacted litter.

When your ecobrick is full and no more can be stuffed in, it's done. If it can be squeezed by more than 10 per cent with one hand, more waste can be added in. You want it to be no longer squashable at all [4].

Use it as a brick in a building – perfect for insulation in a retaining wall for a polytunnel or growing space, or insulating clay- or mud-brick buildings.

Grow your own garden

A big part of our kitchen is the garden. You don't get more transparency in the food system than growing your food yourself! Getting involved in growing your own also really helps you understand the time it takes to grow plants, how flavours change seasonally, and the different edible parts of a plant.

We have lived in a number of urban and rural places with very different garden sizes, from small, shaded balconies to 1-acre market gardens. Size doesn't matter!

Getting started with a garden can be daunting as well as exciting. Our most important piece of advice is ... grow what you would like to eat! Here are some top tips on how to do that.

Plant type

The two main plant types are annual and perennial. Annuals grow for one season, while perennials live for multiple growing seasons. We have a mix in the garden to keep the soil, pollinators and us happy throughout the year.

Annual: kale, carrots, beetroot, cauliflower
Perennial: sage, Jerusalem artichokes, asparagus

Our favourite things to grow in our kitchen garden are herbs, cut-and-come-agains (kale) or plants that give multiple fruits (tomatoes). These are plants that you harvest from and then the plant regrows and you can harvest again, which means you have a continuous return on your investment, and maximize on any growing space.

Bees are what enable many of our plants to fruit so grow bee-friendly flowers and herbs for the pollinators alongside your crops (choose edible ones like borage, dahlias and lavender).

Seeds

We primarily use open-pollinated seeds, where the plants produce seeds with the same characteristics that you can save year on year, and choose organic where possible.

Soil

Soil is the food medium that enables your plants to grow. It's composition, structure, moisture and pH will affect how your plant develops. If planting in pots, you can control this a little easier with compost. For planting seeds we use seed compost, then pot into multipurpose. We avoid peat, due to the damage peat moss mining has caused environmentally. Peat bogs are the largest global terrestrial carbon store.

The most closed-loop way to make soil is to make compost from your food and brown waste scraps (see page 165). Play around with different plants to see what works for you, and you can add compost to your own soil to help your plant get what it needs. We like planting outdoors into soil rather than pots, as the natural ecosystem regulates the plant's growth. For plants that need a lot of feed we add some organic plant food, often seaweed based.

CUT FLOWERS

We started off as a supper club, running interactive sustainability-focused multi-course meals. In decorating a supper club, the norm is often to have cut flowers on the table, or to decorate with vegetation. We've always opted for potted plants, picking wildflowers or aesthetic vegetation from our garden or the farm; think bushels of wheat, hops, gourds and glass gem corn in autumn, or branches of spring blossom, summer flowers, grasses or leaves.

It is lovely to have a big bunch of flowers in the kitchen or dining area, but these don't come without a considerable, and avoidable, environmental cost. Many flowers in the UK are imported from South Africa and Kenya, where they are likely grown in greenhouses with pesticides, fertilizers and herbicides. They are also likely to be air freighted and flown in and refrigerated, in the case of the UK due to their perishability and fragility. And they need a lot of water and land!

We try to keep exotic or imported flower buying to specific occasions where it feels appropriate or traditional, such as get well soon visits. For birthdays, dinner parties and decorating our own kitchens we often have more time to think of an alternative with a less negative environmental footprint and pick a roadside posy or nab something from the garden.

Space and light

Choosing the right plants for your space is important. For larger areas you can choose spreading plants like squash, and for smaller ones choose herbs, climbers (like peas) or cut-and-come-agains (salad, spinach or chard).

Just make sure your plants get adequate light, air flow (especially important indoors), water and feeding. We find we get more pest problems with edible plants indoors than outdoors. Plants will tell you if they are unhappy, and you may have to revisit your growing technique if they wilt, turn yellow or attract pests.

Tomatoes, cucumbers and peppers prefer warmth and a lot of light to give sweet fruit, whereas chard, spinach and brassicas will happily grow in low light levels. Access to a south-facing sun trap is good for young plants, as they need a lot of light to grow, otherwise they bend to the light and end up with weak, drooping stems. Be aware of winter window drafts, freezing temperatures, sun scorch through glass and central heating.

Remember, many plants are insect pollinated, so will have to be put outside to fertilize and produce fruits.

Pests and disease

We net our young plants to prevent rabbits, deer and mice chomping them, particularly in the winter when animals are hungry. We keep nets on brassicas to prevent white cabbage butterflies. You often get powdery mildew on cucurbit leaves (for example, squash, cucumber, courgette and pumpkin) in late summer (our friend Maisie has an amazing technique of spraying them with milk, which kills the fungus!). We also plant decoy plants that attract pests away from our crops (clover for slugs and a few wheat seeds planted in the garden to deter the cat next door). It's best to check online for the particular pests in your area.

Make Your Own: Homemade Garlic Soap Pesticide

This handy little number will help keep pests at bay – it works well on bugs like blackfly. Garlic can kill or deter your pests (depending on the critter), and biodegradable soap disrupts their cell membranes. Unlike conventional pesticide, there is nothing in this spray that will do you or your household harm.

The best way to use this spray on indoor plants is to take the plants outside and spray liberally, making sure to coat the underside of the leaves. Leave to dry and then bring indoors, otherwise you end up with a very pungent garlic-smelling room! Outdoor plants can just be sprayed liberally as needed.

Makes 200 ml (7 fl oz)

Materials

1 whole garlic head, cloves peeled
200 ml (7 fl oz) boiling water
2 tsp biodegradable soap (such as castile)
Spray bottle

Method

Blitz the garlic in a food processor or blender, then transfer to a bowl with the boiling water. Leave to infuse for a day.

Strain the mix, add the soap and pour into a spray bottle. Store in the fridge.

Cooking foundations

In addition to recipes, we wanted to give you a little introduction to the cooking foundations of The Sustainable Food Story kitchen. Some books, chefs and techniques have greatly contributed to our approach to sustainability in the kitchen. Starting from the basics isn't going to be everyone's cup of tea, but if you've picked up this book it probably means you're already well invested in changing your relationship with one of the most important workspaces in your life. By having different tricks up our sleeve for utilizing ingredients, it means we can do more with even the simplest of them. The more you know about cooking and different preparation methods, the more power you have to make choices in line with how you want to be eating. For the health of people and planet, this is generally understood to mean incorporating more wholefoods and plant-heavy options. This is why we focus on foundations that are mostly relevant to doing more with plant based foods.

Sprouting Seeds

We've enlisted the sprouting property of seeds, grains and pulses for a number of our events and supper clubs. Sprouting is a great way to add diversity to your diet. It's also a fun project to teach kids about plants, the connection between the foods we eat and nature's biological processes. Sprouting also gives us another option for fresh, crunchy greens when there's an abundance of root vegetables at the farmers' market, but not yet many fresh greens, or during the hungry gap in early Spring.

To sprout your own seeds, pulses or grains, you don't need any fancy equipment. Alfalfa seeds are a great seed to start with, but try anything you like. They add a nutritious crunch to salads, on top of soups or in sandwiches.

Materials

2–3 tbsp alfalfa seeds
Jar
Piece of muslin or cheesecloth, or even upcycle a pair of washed, torn tights or gauze of some kind
Rubber band or string

Method

Place the alfalfa seeds in the jar and add enough water to cover the seeds by about 1–2 cm (½–1 inch).

Cut out a disc of muslin, cheesecloth, or whatever you are using that is roughly 2 cm (1 inch) wider in circumference than the jar. Place this disc over the jar and secure it with a tight elastic band or string.

After a few hours of soaking, pour the excess water off, straining through the porous cloth.

Shake the damp seeds back so they lie along the side of the jar and leave somewhere you'll see them each day. We usually leave them near the kitchen sink.

Rinse the seeds twice a day for 2–3 days until the seeds have sprouted little white and green tails. Now they are ready to eat.

* ***NOTE:*** *for more information on what the hungry gap is, see here https://wickedleeks.riverford.co.uk/features/local-sourcing-news-farm/what-hungry-gap*

Preserving: Blackberry, Soy and Basil Shrub

Preserving is one of our favourite ways to eat certain fruits and vegetables all year round. It is also a sure fire way to add some pizzazz to your dishes; whether it's homemade strawberry jam to sweeten your peanut butter on toast, preserved lemons to flavour bomb your tagine or condiments for a tangy addition to any dish.

The main home preservation methods involve adding sugar, salt and/or acid which all are natural preservatives. Fermenting, drying and freezing are the other main preserving methods we enlist in our kitchens. See the method for sterilizing jars on page 55.

This recipe is great for using the early autumn glut of blackberries and end of season basil. We also added some green tomatoes that hadn't ripened in time. A shrub is also known as a drinking vinegar, but with the soy sauce addition we like it as a zesty addition to tomatoes or added to sparkling water for a tangy, sweet-savoury drink.

Serves 4

Ingredients

200 g (1½ cups) blackberries
200 g (1 cup) unrefined sugar
1½ tbsp soy sauce
Handful of basil leaves (they can be a bit wilted)
100 g (3½ oz) green tomatoes
200 ml (generous ¾ cup) apple cider vinegar

Method

Wash the blackberries and toss in the sugar and soy sauce. Leave in the fridge overnight. The next day, mix the blackberries with the basil, tomatoes and vinegar in a kilner or glass jar. Leave to infuse for a few days, making sure the liquid covers any solids in the jar.

Sterilize a couple of jars, then strain the liquid through a fine sieve before pouring and sealing it in the jars. If left unopened, the shrub will last for 9–12 months in the fridge. Use within 6 weeks after opening.

TIP:
We like to use the leftover whole blackberries in dishes like the Heritage Tomato and Nasturtium Side Salad on page 148.

Krauting: Cauliflower Leaf Krautchi

This recipe is a hybrid of kimchi and sauerkraut, the delicious probiotic-rich, lactic acid bacteria cabbage ferments associated with Korea and Germany respectively. We add the often-discarded cauliflower leaves to our cabbage ferments. When working in various kitchens as freelance chefs, we'd often end up lugging home a bag of discarded cauliflower leaves if they were in a good condition, knowing we could add them to our picklearium with a few store cupboard ingredients. Chinese leaf, or napa, cabbage is traditional for kimchi, but it is often imported, and we have lots of great local cabbages. White or red firmly packed cabbages make a great krautchi.

Makes 500–600g/
1 lb 1oz–1 lb 5 oz

Ingredients

Cauliflower leaves, from a large, leafy cauliflower
300–400 g (10½–14 oz) cabbage, roughly chopped
4 garlic cloves, peeled and grated
Thumb-sized piece of fresh ginger, peeled and grated
1 tbsp any sugar
1 tbsp gochujang paste or other chilli paste
1 tbsp rice wine vinegar
Sea salt, for the brine

Method

Remove the cauliflower leaves, composting any stems that feel woody, are browning or any damaged sections of the leaves. This is usually the outer two large leaves at most.

Wash and finely chop the cauliflower leaves and stalks. This makes it easier to ensure they release a good amount of moisture and mix well with the salt.

Weigh the leaves and cabbage. Calculate 2 per cent of their weight and add that much salt to the bowl. For example, if you have 500 g (1 lb 2 oz) leaves overall, you will need 10 g (2 teaspoons) salt. This is an easy sum to do, even with a more unwieldy number. If you find yourself mid-cabbage chopping without a calculator handy, or want to try the mental arithmetic, divide by 100 to get 1 per cent and then multiply the answer by two.

Pound and squeeze the cauliflower leaves, cabbage and salt together in a large bowl. We often use our biggest saucepan or a large plastic tub when we're making a huge batch. A flat-ended rolling pin is a useful pounding aid, helping to break down the plant cell walls to release juice from the cabbage leaves. Tenderizing them like this means the juice of the cabbage will form a brine with the salt.

Once lots of juice is welling up at the bottom of the bowl, add in the rest of the ingredients and continue to vigorously mix together.

Transfer to a sterilized airtight container (see page 55) that can easily be 'burped' (as the fermentation process will release carbon dioxide gas), such as a large clip-lock tupperware tub.

The salt brine and cabbage juices should cover the mixture when it is pushed down into the container, so you want it to be a snug fit. Use clean ramekins stacked inside the jar to weigh it down if needed.

Use an outer cabbage leaf to cap off your krautchi, as this piece can be discarded if it's the only part that won't be fully submerged. Try to wedge this scrunched up extra leaf in so that it pushes down your chopped and pounded contents. Alternatively, wrap the leaf around a weight, like a clean stone or small ramekin. This will ensure that even as gas bubbles up throughout the fermentation, your contents stay submerged in the preserving, salty brine.

Leave this mixture to ferment for at least 5 days, but we find 10–15 days is optimal. Remember to burp it every day once the bubbling and fermenting begins. By the end of the process, you will end up with something similar to sauerkraut, but with the punchy flavours of a kimchi, hence 'krautchi'. We love to serve this with eggs on toast for a weekend breakfast, or use in pancakes, like okonomiyaki, in stir-fries or as a gyoza filling.

Tops, tails, stems & skins

There are so many parts of a plant you can eat or use, however, we often cook with a fraction of their edible bounty. If we were to use more of a plant, like the leaves and edible bits that often get thrown away for ease of preparation, it could supplement our food shopping. It means we'll collectively need to buy less food and demand less from our land unnecessarily. If we all wasted less, it would add up to less trips to the shops, less packaging, less waste disposal, less everything – but still just as much food on the table! Here are some less known edible plant parts and tips on what to do with them.

Tops: the tops of leeks can be chopped up and used in the same dishes as the bottom of the allium. Leek tops might need a little longer cooking, but they still taste great. Young, fresh beetroot tops are much like chard and can be used similarly, or as a spinach substitute, or in beetroot leaf borani (search online for ecochef Tom Hunt's recipe). Chopped up hollow courgette stalks look like a green penne pasta, and the very tops are crunchy and edible and can be blitzed into soups. The tops of gone-to-seed brassicas and lettuces you might grow at home, such as broccoli, kale, rocket and mizuna, can also be used in your cooking (flowers can be added to salads, and anything else can be steamed/boiled before using to remove any bitterness).

Stems: We chop broccoli, cauliflower and mushroom stems up small and use them in the same dishes as the tops.

Skins: carrots, parsnips, Jerusalem artichokes, beetroot, potatoes, mushrooms, squash, some pumpkins – the skins of these vegetables are all edible! No need to peel them, just give the veg a wash and they are perfectly good to go. Instead of washing mushrooms, however, we advise dusting any dirt off with a brush or clean, damp cloth. This is because mushrooms absorb water and are easily damaged.

Skins: onion skins can be used for dying, as can beetroot, if you decide they're too knobbly or blemished for eating. You'll just need to follow vegetable dye instructions, and use a household mordant-like salt or vinegar to set the dye afterwards.

Roots: leek and spring onion roots make a delicious garnish and can be cleaned, dried and deep-fried for crispy toppings.

A guide to cooking veg

In terms of sustainability, eating food raw of course uses the least energy, but isn't always the best preparation method. Employing a variety of cooking methods to impart different flavours, textures and desirable levels of tenderness to vegetables, grains and pulses is probably the most important way cooking methods contribute to a more sustainable kitchen.

If we think of vegetables as being a boring, boiled side, it's no wonder meat, eggs, cheese and creamy sauces have so often taken centre stage.

Here are our preferred methods to prepare a range of our favourite veg – those commonly available at our local farmers' markets.

HOW TO PREPARE YOUR VEGETABLES

	Steam	Boil	Roast	Fry	Raw	Grill	Griddle	Deep-fry	BBQ
Asparagus	•	•	•	•	•	•	•		•
Aubergine			•	•		•	•	•	•
Artichoke, Jerusalem		•	•	•	•			•	
Beetroot	•	•	•	•	•				•
Broad beans	•	•			•				
Broccoli	•	•	•	•	•				•
Brussels sprouts	•	•	•	•	•				•
Cabbage	•	•		•	•		•		•
Celeriac	•	•	•	•	•	•			•
Carrot	•	•	•	•	•	•			•
Courgette			•	•	•	•	•	•	•
Cucumber					•				
Fennel			•	•	•		•		•
Kale	•	•	•	•	•				
Leek			•	•			•	•	•
Mushroom			•	•	•	•	•	•	•
Onion			•	•	•		•		•
Parsnip	•	•	•		•			•	
Peas	•	•			•				
Pepper			•	•	•	•	•		•
Potato	•	•	•	•				•	
Radish			•	•	•		•		•
Runner beans	•	•			•	•			
Squash/pumpkin	•	•	•						•
Swede	•	•	•		•				•
Sweetcorn	•	•	•	•		•			•
Spinach	•	•		•	•				
Spring onion			•	•	•		•		•
Tomato					•	•			•
Turnip	•	•	•		•				

Cooking methods

Microwave

Having grown up without a microwave, Sadhbh is a recent microwave convert. Energy requirements are reduced by as much as 80 per cent when using a microwave for cooking or reheating small portions, instead of an oven or stove.

Some of our top microwave uses include:
- Baking quick, individual mug cakes, when you need a sweet treat but don't want to have a whole cake in the house.
- For semi-cooking a baked potato – we pierce the potato and microwave on high for 6–10 minutes before finishing in the oven, vastly reducing the oven cooking time.
- Reheating leftovers – reheat leftovers straight in the bowl you stored them in. This also saves water due to less washing up.

The downside of microwaves is of course that they are also associated with increased consumption of ready meals, and are another gadget to buy. If you don't already have one and have managed without one for this long, it might not be worth going out and purchasing one unless you plan to use it a lot. A combi-oven provides the best of both, with microwave and oven functions, making it easy to both reheat and freshly cook.

Is induction cooking the future?

Many kitchen providers are advising installation of induction stoves/hobs. These are also becoming more commonplace in professional kitchens. Although pans made from non-magnetic materials are rendered unusable on induction hobs – induction is more energy efficient than the alternative electric element or gas burner heating. Induction heats only the pot, and not the air surrounding it. It's also seen as safer, as only the pot and food get very hot, but the ceramic surface above the induction coil stays relatively cool. This means the surface is cool enough to wipe down more instantaneously and these hobs are less likely to cause kitchen fires.

Salt and salting

There are a number of sources for salty, umami flavour, be that conventional salt, yeast extracts, soy sauce, ferments or cheese. Salt flavour can be present without just adding salt. We have been inspired by Samin Nosrat's focus on salt and its importance as an enhancer of your food's natural flavour. We opt for sea salt as it is obtained from evaporated sea water whereas rock salt is mined. Our favourite salt in the UK is Halen Môn who have won a Queen's Award for sustainability. It's important to taste the salt you are cooking with and get to know it, as some salts are 'saltier' than others, some have larger crystals and so on.

Salting should be in the foundations of your dish. 'How much salt do I add?' is a question we get asked a lot. When boiling it should be proportional to the amount of time your produce is in the pot. Short cooking times should equal more salt. If you are trying to decide whether to add salt to say, a stew, then taste the dish and decide whether it tastes of enough. Sometimes you just need to add a couple more pinches and the dish will pop.

Cooking

PLANTS *(see pages 83–112)*
Plants live at the core of our cooking. There is increasing population-level research to show that a diet consisting predominantly of plants and wholefoods has a lower environmental impact than a diet high in animal products. There are a growing number of studies showing that it is better for general population health.

When choosing a recipe or planning a menu our first port of call is to think about what fruit and vegetables are in season. Fruit and veg are nutritional powerhouses packed with vitamins, minerals and fibre essential for general health, and the more recently understood gut microbiome diversity. Our simple rule of thumb is to make your plate as colourful as possible, which is easily achieved when including a diversity of fruit and veg.

Weirdly, the likes of bread and vegetables have been relegated to being side dishes; hopefully the recipes in this book will help you put plants back at centre stage. For the food we serve to be more sustainable, this needs to become the norm.

VAGUEAN *(see pages 113–127)*
Our chef friend, Mickey, introduced us to the term vaguean. She ate mostly vegan, but if something with animal products was going to be wasted, or a recipe was too tricky to adapt and there was dairy or eggs to be used in the kitchen, she wasn't going to berate herself for not being 100 per cent vegan all the time. We do endeavour to adapt recipes to have plant alternatives, but

some classic dishes don't work out the same, so we do use eggs and dairy, too, on occasion. When we do use these ingredients, we make sure they are produced in the best possible way.

This also helps us avoid having to take B12 supplements, and often avoids wasting things that might be in kitchens we're working in that need taking home at the end of service. As food-pun enthusiasts, we thought the portmanteau vaguean (from vaguely vegan) fits this section of recipes pretty well. It's pretty much flexitarian, or reducetarian-style cooking, and the recipes are vegetarian, but often adaptable to be vegan.

OMNIVORE *(see pages 128–137)*
We don't cook very much meat and fish in our own kitchens, however, sometimes we have a request for something omnivore, or our bodies tell us maybe we are in need of a little iron or B12. This is especially true being women, with a greater requirement for iron; the more readily absorbed heme iron is only found in animal products. We are also conscious that if we eat dairy we should really use the byproducts of the industry – eating goat's cheese means we should be eating the billy goats, for example.

With waste in mind, we also try to focus on under utilized animal products such as blood and offal. These are our go-to meat or fish choices, with environmental sustainability relevant to our locale in mind. We use them as an opportunity to tell a story about what sustainability-focused agriculture and aquaculture could look like.

SAUCES AND SIDES *(see pages 138–151)*
We've included some of our favourite sauces that we've adapted and developed over the years, but learning the basics of sauces

will stand you in good stead on a journey towards a more sustainable kitchen.

Sauces are something we get asked about a lot, as they are the basis of so many dishes. Although most home cooks and cafe chefs don't have much reason to make a velouté or an espagnole these days, it's the concepts behind them that will help you understand the chemistry of cooking. And the more you practice things, and understand the building blocks, the more confident you will be to adapt traditionally meat or animal product dishes into more plant-based versions. If you can hit the sweet spot of getting great flavour and texture with little effort, and confidence in the kitchen, you'll also be less likely to produce dishes that you're tempted to throw in the compost. Most things can be rectified, so believe in yourself, and get comfortable with the basics first.

We have also included some of our favourite sides, which can be made as an addition to a dish, side to a meal or a snack when you're out and about.

RECIPE NOTES:

- All herbs are fresh unless stated otherwise.
- Vegetables are only peeled if necessary (onions), most skins are left on (carrots, butternut squash for example) unless stated otherwise.
- All oil used is rapeseed or vegetable unless stated otherwise.

Giorgia's Leftover Grain Seeded Sourdough

This super seeded bread using leftover grains was first brought to the Skip Garden kitchen menu by intern-turned-community-chef-extraordinaire Giorgia Lauri. We all met through our connection to the Skip Garden, so it's nice to think of this forgiving fusion bread, accepting all wholesome seeds, grains and nutty flours around, as somewhat representative of the collective that is The Sustainable Food Story. Bake one for you and one for a friend. For our Sourdough Starter recipe, see page 151.

Prep: 30 minutes, starting day before to allow for seed soaking time and overnight ferment. Start 5 days before if you're starting your own sourdough starter from scratch.
Cook: 1 hour 15 minutes

Makes 2 small loaves

Ingredients

260 g (2 cups) mixed seeds (such as sunflower seeds, pumpkin seeds, sesame seeds, linseeds)
480 g (1 lb) sourdough starter
300 g (10½ oz) cooked and cooled grains (such as barley, rice, spelt, rye, oat groats)
280 g (2¼ cups) rye flour
230 g (scant 2 cups) spelt flour
2 tbsp treacle
3–4 level tsp (22 g/1 oz) fine sea salt

Method

In a large mixing bowl, cover the seeds with 300 ml (generous 1¼ cups) water and set aside to soak for a few hours.

Once the seeds have had their soaking time, add the rest of the ingredients to the bowl along with 80 ml (⅓ cup) water. Stir thoroughly until everything is evenly combined.

Cover with a bowl cover (see page 53) and leave in a warm, draft-free place to ferment and rise for 1 hour. Transfer to the fridge to rest overnight.

The next day, remove from the fridge and divide between two 450 g (1 lb) loaf tins, and leave for about 2 hours to come up to room temperature.

Preheat the oven to 220°C (200°C fan/425°F/Gas 7).

Place both loaves in the oven and bake for 15 minutes, then reduce the temperature to 200°C (180°C fan/400°F/Gas 6) and bake for 1 hour.

To check if the loaves are cooked, knock on the bottom of the loaf – it should sound somewhat hollow. Leave the loaves to cool in their tins before slicing.

These loaves should last for over a week in a paper bag or bread bin. If you want to keep both loaves, pre-slice the second loaf and put it in the freezer so you have a nutritionally diverse toast option always ready to go (tahini and marmite make an excellent topping combination).

Shortcrust Pastry

Contrary to popular belief, making a good shortcrust is both easy and satisfying (Abi Aspen would have contested this before we started cooking together!). It can be the base for so many great dishes, be it a cosy night in with a pie or wanting to wow your friends with a fancy tart. It's a staple recipe in the TSFS kitchen. Shortcrust is also forgiving in the flour you can use, so you can freely choose your ingredients for sustainability and flavour.

Prep: 15 minutes
Cook: 30 minutes

Makes enough for a 23 cm (9 inch) tart case

Ingredients

300 g (2⅓ cups) wholemeal flour (or flour of choice), chilled
150 g (5½ oz) vegan block (or butter), chilled and diced
1–2 tbsp ice-cold water
Fine sea salt

Method

Place the flour in a bowl, sprinkle in a pinch of salt and mix through evenly. Add the block (or butter) and rub into the flour using your fingertips to keep the mixture cool **[1]**. Carry on until it resembles fine breadcrumbs. Alternatively, you can use a food processor, pulsing gently so it doesn't get too hot.

Bring the mixture together into a firm dough **[2]**. Start by adding 1 tablespoon of the water. If the dough is very crumbly, add the second tablespoon to help it along. Turn out onto a work surface to bring it together, but be careful not to knead it as this encourages gluten to form and can make the pastry tough.

Flatten the pastry into a round disc **[3 and 4]** and wrap in a beeswax wrap (see page 51) or alternative. Place in the fridge and leave to chill for 30–60 minutes before using or freezing. The chilling process allows the gluten to relax and prevents the pastry from shrinking too much when it cooks.

1

2

3

4

Chickenish Pie

This is a vegan take on a classic creamy chicken and veg pie topped with pastry. Puff pastry works for topping this pie, but as shop-bought puff pastry often contains palm oil and no wholemeal flour, we switched to topping it with homemade wholemeal shortcrust. All of the separate components (pastry, white sauce, filling) can be prepared in advance, meaning the pie can be ready within 30 minutes for a plant-based Sunday lunch.

Prep: 40 minutes
Cook: 30 minutes

Serves 4–6

Ingredients

2 tbsp oil
2 leeks, sliced into rounds (including green tops)
1–2 carrots, diced
1–2 celery stalks, or half a celeriac, diced
150 g (5½ oz) mushrooms, sliced
4 garlic cloves, finely chopped
280 g (10 oz) rehydrated TVP chunks, or other chicken-like protein substitute
1 quantity of Vegan White Sauce (see page 139)
5 thyme sprigs, leaves removed
½ quantity of Shortcrust Pastry (see page 84)
1 tbsp plant-based milk, such as soya, for glazing
Sea salt and ground black pepper

Method

Preheat the oven to 220°C (200°C fan/425°F/Gas 7).

Heat the oil in a deep, wide pan over a low heat. Add the leeks, carrots, celery or celeriac, mushrooms and garlic, stir and put the lid on. Cook for 5–8 minutes until vegetables are soft. Stir occasionally as you don't want them to catch and brown. This umami mirepoix is the vegetable pie base. Add the TVP pieces and cook for 2 minutes.

Add the white sauce to the veg mixture. Season with salt and pepper and add the thyme leaves. Heat through for another 4–5 minutes, making sure the white sauce is coating everything. Pour this pie filling into a large pie dish.

Roll out the pastry on a lightly floured surface. Place the pastry over the pie filling, tucking in the sides or trimming off the excess. Brush with a little milk. Excess pastry can be used to make a pattern for the top – we often fashion small leaves or a lattice.

Bake for 20–25 minutes until the top is going golden brown all over and the filling is bubbling. Leave the pie to stand for 5 minutes before serving

Pumpkin Seed Blondie with Seasonal Fruit

This recipe conjures up fond memories of cooking for 100 people per night at Shambala Festival with our comrade-in-arms, Sarah, wearing nothing but underwear and aprons and then dancing until 1am. Food is an important vehicle to take you back to those moments. Our blondie started life as a frangipane, which we both love. However, trying to diversify our almond and dairy consumption inspired us to try some alternative ingredients. We ended up using pumpkin seeds and some aquafaba leftover from making a salad. Guess what … it worked! So we came up with a new hybrid – an environmentally conscious, gooey frangi-blondie.

Prep: 15 minutes
Cook: 30 minutes

Makes 1 x 25 cm (10 inch) diameter tart

Ingredients

- 200 g (scant 1½ cups) pumpkin seeds
- 170 g (generous ¾ cup) light brown sugar
- 200 g (1½ cups) plain flour
- Zest of 3 unwaxed lemons
- ½ tsp bicarbonate of soda
- ½ tsp ground cardamom
- 120 ml (½ cup) vegetable oil
- 140 ml (generous ½ cup) aquafaba
- 200g (7 oz) seasonal fruit. We like plums, pears or apples cut into eighths, or sticks of rhubarb, trimmed and chopped into 4 cm (1½ inch) pieces on the diagonal and tossed in sugar.

Method

Preheat the oven to 200°C (180°C Fan/400°F/Gas 6). Line a 25 cm (10 inch) tart tin with greaseproof paper.

Put the pumpkin seeds in a food processor or blender and pulse until they resemble ground almonds.

In a large mixing bowl, stir together the sugar, flour, lemon zest, bicarbonate of soda, cardamom and ground pumpkin seeds. In a separate bowl or jug, lightly whisk the oil and aquafaba together until combined (but not frothy), then pour into the dry ingredients and stir until you have a thick, gloopy mixture.

Spread the mix into the lined tart tin evenly. Layer the fruit in a pattern of your choice on top, gently pressing it into the frangipane. Bake for 20–30 minutes, or until the top has a light brown crust.

Smokey Breakfast Porridge with Melted Leeks

We first served this breakfast on a summer retreat in the English countryside for the lovely women at Ladyshift (a collective running wellness event for women). This umami-packed, savoury breakfast was an easy-to-make alternative to a sweet porridge in our outdoor kitchen, plus it's a great chance to use the top green part of the leek, which is often removed unnecessarily. We love the outdoors (you will often find us climbing rocks or camping) and cooking outside and luckily this recipe is great for cooking over the campfire.

Prep: 15 minutes
Cook: 30–40 minutes

Serves 4

Ingredients

4 tbsp smoked rapeseed oil
1 leek, finely chopped (including green tops)
2 garlic cloves, finely chopped
200 g (⅓ cup) oats (we use steel-cut oats as it gives the porridge good texture)
1 litre (4 cups) oat milk (or any milk desired)
250 ml (1 cup) oat cream
2 tbsp bouillon
2 tsp red miso
1 tsp yeast extract (such as Marmite)
1 tsp soy sauce
½ tsp dried mushroom powder (optional)
1 tsp maple syrup (optional)
Handful of leafy greens (such as kale or spinach)

Method

Heat the smoked oil in a large saucepan over a medium heat for 2 minutes. Add the leeks and cook for about 5 minutes until softened. Add the garlic and cook until the aromatic garlic smell begins to appear, stirring often so the vegetables don't catch.

Add the oats and coat well in the oily mixture. Pour in the oat milk and oat cream and stir. Add the bouillon, red miso, yeast extract, soy sauce, dried mushroom powder and maple syrup, if using.

Keep stirring gently over a low heat for 20–30 minutes until the oats have soaked up most of the liquid leaving a smooth porridge. Add the leafy greens for the last 5 minutes and let them wilt into the porridge. You can test to see if the porridge is ready by trying a little taster of oat – if it is soft, then the porridge is ready to eat.

Heritage-grain Gyoza Stuffed with Stir-fry Leftovers

Many of our friends order gyoza as a takeaway dish, one of those things that seem difficult and fiddly to make. However, the process is quite simple – we even make them with children in cookery classes! A great way to use up leftover stir-fry (alternatively use the filling recipe below), these are an easy midweek meal or a show-stopping dish for dinner with friends. We like giving the dish a twist by using local heritage grains for the wrappers.

Prep: 1 hour, plus resting
Cook: 20 minutes

Serves 4

Ingredients

For the gyoza wrappers

125 g (scant 1 cup) heritage-grain wholemeal flour
125 g (1 scant cup) strong white flour, plus extra for rolling
Pinch of sea salt

For the stir-fry

3 tbsp sesame oil
4 garlic cloves, crushed
1 tbsp peeled and freshly grated ginger
½ small mild red chilli, finely chopped (or 1 tsp dried chilli flakes)
4 spring onions, chopped
2 carrots, cut into matchsticks
5–6 mushrooms, thinly sliced
⅙ of a small white cabbage, core removed, leaves thinly sliced
1 tbsp soy sauce

Method

For the gyoza wrappers, measure out the flours and salt into a large mixing bowl and combine. Slowly add 150 ml–175ml (⅔–¾ cup) water, incorporating a little at a time. Add enough water to form a moist dough, but douse in flour if it feels too sticky. Knead the dough for a few minutes, adding large pinches of flour if needed, to reduce stickiness as the flour is absorbed. Cover the dough ball in its bowl and leave to rest for at least 1 hour.

Once the dough has rested, divide it into roughly 40 smaller dough balls, about 10–13 g (⅓–½ oz) each. Roll each piece of dough into a circle, flouring the surface and rolling pin regularly to prevent sticking. The wrappers can be stacked on top of each other for up to 20 minutes, if well doused in flour so they don't stick together. If it will take you longer than that, lay them out on floured trays or boards, or have someone else filling them straightaway.

For the stir-fry, heat 2 tablespoons of the sesame oil in a wok or large frying pan and add all the ingredients, except the soy sauce. Stir-fry everything together over a high heat for about 5 minutes until the carrot is starting to soften and the mushrooms are browning. Add the soy sauce and fry for a minute more, stirring well. Set aside to cool before assembling the gyoza.

Once the stir-fry has cooled, spoon 1½–2 teaspoons of the stir-fry mixture into each wrapper [1]. With your fingers or a pastry brush, brush one edge of each wrapper with a little water, and pleat one side before sticking on top of the opposite side, sealing in the filling [2]. Pinch each one tightly closed to make sure it doesn't leak or break open when frying [3]. Repeat with the rest of the wrappers, laying each gyoza on a lightly floured tray.

[method continued overleaf]

1

2

3

4

To cook the gyoza, heat the remaining tablespoon of sesame oil in a non-stick frying pan that has a lid or cover. Place as many gyoza as will comfortably fit in the pan into the hot oil, with the flat side down. Fry for 3–4 minutes over a medium-high heat until the underside of each dumpling is golden brown.

Cover three-quarters of the pan with its lid, and pour in enough water to submerge the bottom one-quarter of the gyoza –150–200 ml (⅔–¾ cup) should be enough. Water and hot oil don't like each other, so it will spit and steam; keep your hand well back and the lid as tightly over as you can.

Cover the pan fully with the lid and leave to steam for another 4–5 minutes to cook the dough. Once the water has evaporated, the base of the frying pan is dry and the dumpling bases have crisped up again, your gyoza are ready to serve.

Repeat this frying step as many times as you need to cook all the gyoza. You may need to wipe out the pan between batches if the oil is burning.

We like to serve these with a sweet chilli dipping sauce or a small dish of soy sauce, sauce as a snack or starter, or with broccoli dressed in chilli and garlic oil as a main. **[4]**

TIP:

Gyoza are ideal for freezing as they can be cooked from frozen. Place uncooked gyoza in the freezer on a lined baking tray, spread out in a single layer. Once frozen, they can be bundled together in a resealable freezer bag or a tub. If cooking from frozen, add enough water to come halfway up the side of the dumpling when steaming, following the same steps as cooking from fresh, above.

Miso Mushrooms

Mushrooms are pretty badass, adding texture and earthy tones to any dish. They also soak up flavour exceedingly well. This technique of marinating mushrooms adds loads of umami and depth of flavour, and is inspired by a recipe we made for a kids' cookery class that covered different ways of bringing intense flavour to vegan cooking. These are great for topping the Speltotto for Every Season (see page 122) or having with grains, tofu, beans and veg for a simple one-bowl dinner.

Prep: 35 minutes
Cook: 10 minutes

Ingredients

- 3 tbsp balsamic vinegar
- 3 tbsp oil
- 2 tbsp miso
- 2 garlic cloves, very finely chopped
- 2 tsp wholegrain mustard
- 4 large portobello mushrooms, or 250 g (9 oz) any large mushrooms, like chestnut mushrooms
- Sea salt and ground black pepper

Method

To marinate the mushrooms, combine the balsamic vinegar, oil, miso, garlic, mustard, salt and pepper to taste in a shallow dish or roasting tin. Clean and trim the mushrooms if needed. If the mushrooms have been kept for too long and have started to go a bit slimy, they're generally fine to eat once the skin of the cap has been removed and they've been cooked.

Place the mushrooms into the marinade, spreading it all over and getting it into the underside of the cap. Leave to marinate for 20–30 minutes, turning and re-coating in marinade halfway through.

Preheat your grill to high (alternatively, these can be baked in the oven at 200°C/180°C fan/400°F/Gas 6 if you don't have a grill, or would be great barbecued). Place the mushrooms bottoms up on a roasting tray, and tip the pan so that all the marinade gathers in one corner. Spoon the excess marinade into the upturned caps. Place under the grill, and grill on high for about 5 minutes. Keep an eye on them; they should start to get a nice char.

Flip the mushrooms over and grill for another few minutes until the tops are wrinkling, and they ook as if they've softened to a nice tenderness. Slice larger portobellos before serving and fan them out on a plate or bowl, but leave smaller mushrooms whole.

Cumin, Sesame and Cauliflower Soup with Crispy Cauliflower Leaves

We first had something similar to this soup in Delhi, and were bowled over by its flavour combination and creaminess. It's hardly surprising that this vegan belly-warmer was found in a cafe in Delhi, considering India is where some of the world's most popular vegan dishes originate.

Prep: 10 minutes
Cook: 30 minutes

Serves 4–6

Ingredients

- 1 large cauliflower, broken into florets, woody end of base removed and leaves reserved
- 3 tbsp toasted sesame oil, plus extra for drizzling
- 2 tbsp cumin seeds, plus extra to garnish
- 1 large onion, chopped
- 4 garlic cloves, chopped
- 600 ml (scant 2½ cups) vegetable stock
- 3 tbsp tahini
- Sea salt and ground black pepper
- 2 tsp toasted sesame seeds, to serve

Method

Preheat the oven to 200°C (180°C fan/400°F/Gas 6).

In a large roasting tin, toss the cauliflower florets in 2 tablespoons of the toasted sesame oil, with the cumin seeds and some salt and pepper. Roast for 20–30 minutes. Add the chopped cauliflower leaves halfway through to get them crispy.

Meanwhile, the remaining tablespoon of oil in a large saucepan and sauté the onions and garlic over a low heat for 10 minutes.

Once the cauliflower is nicely roasted, add the florets (but keep back the leaves) to the onion mixture along with the vegetable stock and tahini. Transfer to a food processor or blender (or use a handheld blender) and blitz until smooth. Season to taste.

We serve this soup topped with the reserved cauliflower leaves, a drizzle of toasted sesame oil, toasted sesame seeds and an extra pinch of cumin to garnish.

Mushroom and Tofu Burgers

Focussing on grains and pulses as the plant-based protein source in a meal meant we hadn't used tofu in our cooking much until recently. We are aware that unless you make your own (that's dedication!) it often comes in plastic packaging. However, it is a great gateway to eating more plant-based foods, so we do make the exception and buy it for certain dishes. A lot of tofu brands have in-depth research into their carbon footprints and sourcing on their websites. Knowing this reassured us that the soya for tofu isn't usually rainforest-sourced. This burger mix is really useful and can be used for some excellent meatballs as well (see Note below).

Prep: 15 minutes
Cook: 10–15 minutes

Serves 4

Ingredients

- 2 tbsp oil, plus extra for frying
- 1 onion, roughly chopped
- 3 garlic cloves, finely chopped
- 250 g (9 oz) mushrooms, sliced
- 1 tsp mixed dried Italian herbs
- 1 tsp sweet smoked paprika
- 2 tsp miso
- 3–4 tsp veggie Worcestershire sauce or soy sauce
- 2 tbsp ground flaxseeds
- 1 tbsp nutritional yeast (optional)
- Block of firm tofu, drained and roughly chopped (350–400 g/12½–14 oz)
- Chickpea flour, as needed
- Handful of breadcrumbs (optional, if the mixture seems wet and you have them to use)
- Lettuce, tomato slices, pickles, sauces and buns, to serve

Method

Heat the oil a large pan over a medium heat. Fry the onion, garlic and mushrooms until most of the water has evaporated from the mushrooms and they're starting to brown. Mix in the dried herbs, paprika, miso, Worcestershire sauce or soy sauce, flaxseeds and nutritional yeast, if using.

Stir together for a few minutes. Add a little water if it's too dry, or add chickpea flour if it's too wet and not starting to stick together as you combine.

Stir in the tofu and mash together. We use a potato masher, which slices through the tofu and cooked mushrooms, combining them into a protein-full, umami texture. At this stage you can add a handful of breadcrumbs to help combine if the mixture seems wet; this is dependent on different firmness of tofu brands. Breadcrumbs also help to bulk the mixture out, and it uses up stale bread if you have it. Leave the mixture to cool for a few minutes.

Shape the mixture into burgers with your hands, pressing firmly together so they won't crumble when you flip them. In a large frying pan, heat a little oil over a medium heat and shallow-fry the burgers for about 10 minutes until browned and crisp all over.

Serve the burgers with lettuce, tomato slices, pickles and sauces of your choice in a seeded, wholemeal bun.

* ***NOTE:*** *to make meatballs instead of burgers, make the mixture, but roll into balls no bigger than golf-ball sized, so they have more surface area for crispiness. Shallow fry in hot oil in a large frying pan over a medium heat for about 10 minutes until browned and crisp all over. You may need to do this in batches, depending on pan size. Serve in the Tomato Sauce Base on page 138 with pasta or grains of your choice.*

One-pot Bowl for Every Season

We wanted to share some of our favourite recipes for one-pot meals, using relatively low-cost seasonal veg and pulses or grains. One-pot dishes are generally energy- and time-efficient, and produce less washing up. Win win for you and your kitchen's environmental impact. By using ingredients seasonal to where you live for go-to recipes like these, you'll have as sustainable as possible a dish in your back pocket for an easy, nutritious meal.

Spring: Spring Greens and Quinoa Stew

Prep: 25 minutes
Cook: 35 minutes

Serves 4

This dish is inspired by an Italian vignole, or spring stew, crossed with our favourite spring greens quinoa risotto by Hemsley + Hemsley, and a Peruvian quinoa soup our friend Lisa, one of our first fab interns at the Skip Garden, introduced us to. We use quinoa as this nutritious protein-rich pseudograin grows well in temperate climates. There are great regional growers providing an alternative to importing it from South America. In the UK, British organic quinoa can be found in many health-food shops and larger supermarkets.

Ingredients

3 tbsp oil, plus extra for drizzling
3 spring onions or 1 small leek, sliced
3 garlic cloves, crushed
500 ml (2 cups) vegetable stock
100 g (½ cup) quinoa
1 fennel bulb, halved and thinly sliced (optional; may still be in season late winter/early spring)
120 g (1 cup) or a few stalks of purple sprouting broccoli or spring greens, sliced
Bunch of chard, roughly chopped, stalks separated
240 g (2 cups) broad beans (frozen if not quite in season yet)
240 g (2 cups) peas (frozen if not quite in season yet)
40 g (⅓ cup) mint leaves, chopped
30 g (¼ cup) parsley, chopped
Sea salt and ground black pepper
4 lemon wedges, to serve

Method

Heat the oil in a large saucepan over a medium heat and once hot add the spring onions or leek and the garlic and cook for 3–4 minutes. Stir in the stock and quinoa and cook for 8 minutes.

Add the fennel, if using, broccoli or spring greens, chard stalks and broad beans and cook for another 5 minutes.

Add the chopped chard leaves and the peas and cook for about 4 minutes until the peas are tender crisp.

Stir in the herbs and season well with salt and pepper. Serve in large bowls with a drizzle of oil and lemon wedges for squeezing.

Summer: Red Pepper and Courgette Tagine

Prep: 30 minutes
Cook: 1 hour

Serves 4–6

Ingredients

2 red onions, halved
Thumb-sized piece of fresh ginger, peeled
6 garlic cloves
2 tbsp oil
2 cinnamon sticks or 1–2 tsp ground cinnamon
2 tsp ground cumin
2 tsp ground coriander
1 tsp ground turmeric
1 tsp sweet smoked paprika
½ tsp cayenne pepper
1 small or ½ large roasted squash, chopped into chunks
1 large carrot, cut into batons
3 red peppers, cut into long wedges
2 courgettes, cut into batons
400 g (14 oz) tin chopped tomatoes
500 ml (2 cups) vegetable stock
400 g (14 oz) tin chickpeas, drained and rinsed
8–10 prunes, dried apricots, figs or dates, roughly chopped
2 small preserved lemons, finely chopped (or juice of 1 lemon)
100 g (¾ cup) toasted nuts, such as agroforestry almonds
1 tbsp honey or agave syrup
1 tbsp harissa paste
Couscous, quinoa or bulgur, to serve
Herbs (such as parsley, mint and coriander), to serve

This recipe comes from the Skip Garden heydays, when community chef Vero had returned from Palestine full of flavour inspiration and touching stories of community bonds over shared meals. She was cooking all sorts of warming stews on the mobile catering bike that used to bring the vibes of the urban growing space to the rest of King's Cross.

Method

First, make a base paste by blending the onions, ginger and garlic in a food processor or blender, or by chopping it all very finely together if you don't have one.

Heat the oil in a large heavy-based saucepan and add the blended paste and all the spices, including the cinnamon sticks. Sauté for 5–10 minutes until the onion paste is sticky and caramelized.

Add the prepared vegetables, chopped tomatoes, stock, chickpeas, dried fruit, preserved lemons, half the nuts, the honey or agave and harissa paste and stir everything together. Bring to the boil, then stir.

Reduce the heat to a simmer and cook for 40–50 minutes with the lid on, stirring occasionally to ensure it's not sticking. Top up with a little water if it gets too thick and starts to stick.

Serve with wholewheat couscous and herbs – we use any combination of parsley, mint and coriander – or other cereals such as quinoa or bulgur, cooked according to their instructions.

TIPS:

- This recipe can be adapted for any mixture of veg, depending on what's in season. Replace the peppers and courgettes with more squash, pumpkin and parsnips in autumn and winter months.
- This slow cooking can also be done in the oven, if you have the right pot for it, or use a slow cooker as the most energy-efficient way to cook until tender. Cook in an oven preheated to 200°C (180°C fan/400°F/Gas 6) for up to an hour, with the lid on.

Autumn: Jerusalem Artichoke and Blue Pea Stew

Prep: 20 minutes
Cook: 4 hours on the hob/
24 hours in the slow cooker

Serves 4

Ingredients

6 tbsp oil
2 onions, thinly sliced
1 leek top, thinly sliced (optional)
4 garlic cloves, crushed
3 tsp smoked paprika
3 tsp mild chilli powder
2 tsp ground coriander
1 tsp ground cumin
8 tbsp red wine vinegar
400 g (14 oz) Jerusalem artichokes, cubed
300 g (1½ cups) dried blue peas, soaked overnight (or 2 x 400 g/14 oz tins, drained)
2 handfuls of kale
2 tsp yeast extract (such as Marmite) or miso
250 ml (1 cup) vegetable stock
2 x 400 g (14 oz) tins plum tomatoes
Handful of freshly chopped parsley, to serve (optional)
4 slices of toasted sourdough, to serve (optional)

Jerusalem artichokes – or 'fartichokes' as we fondly call them due to their high concentration of the soluble fibre inulin, which feeds our good bacteria but makes us toot – are prolific growers with an edible flower that comes back year on year. You can substitute any local pulse for our UK blue peas. This stew is a great hearty dinner to warm the cockles.

Method

Heat 3 tablespoons of the oil in a large saucepan over a medium heat for 2 minutes, then add the onions and leek tops, if using, and cook over a low heat for about 10 minutes until softened. Add the garlic, half the spices and half the red wine vinegar, stir and cook until the onions and leeks are slightly browned and dark red.

Add the remainder of the ingredients to the saucepan and give it a good stir. Leave to cook over a low heat for 4–6 hours on the hob or transfer to a slow cooker for 24 hours. To eat, sprinkle with freshly chopped parsley and serve with toasted sourdough to mop up the sauce.

Prep: 20 minutes
Cook: 20–30 minutes, or overnight in the slow cooker

Serves 4

Ingredients

6 tbsp oil
1 onion, diced
3 garlic cloves, grated
Large thumb-sized piece of fresh ginger, peeled and grated
1 small red chilli, deseeded and diced (optional)
3 tsp yellow mustard seeds
3 tsp ground turmeric
3 tsp ground cumin
3 tsp ground coriander
3 tsp goda masala
400 ml (1⅔ cups) vegetable stock
400 ml (1⅔ cups) coconut milk
400 ml (14 fl oz) tinned chopped tomatoes
Juice of 1 lemon
300 g (1⅔ cups) red lentils
400 g (14 oz) mixed root vegetables (carrots, beetroot, parsnips, turnip), grated
300 g (1½ cups) brown rice (75 g/2¾ oz) per person), rinsed
2 handfuls of cabbage or kale leaves (optional)
Fine sea salt

For the flatbreads:

200 g (scant 1½ cups) wholemeal flour, plus extra for dusting
Pinch of sea salt
2 tbsp oil

To serve:

Small handful of coriander chopped
Pink Pickled Onions (see page 141)
4 radishes, thinly sliced

Winter: Root Veg Dahl with Pink Pickled Onions and Flatbreads

Depending on how decadent and intense you want it to be, this dahl can be a quick 20–minute dish or a slow overnight wonder. We love this both as an easy one-pot dinner or to feed many hungry tummies, and cooking a bit more than you need means you end up with a great lunch the next day. We tend to serve our dahls thick, inspired by Gujurati lachko dal.

Method

Heat the oil in a large saucepan for 2 minutes over a medium heat, then add the onion. Cook for 5 minutes or so, until softened, then add the garlic, ginger and chilli, if using. Continually stir for a couple of minutes, then add the mustard seeds, turmeric, cumin, coriander and goda masala, if using, to toast a little and release their oils. If the mix begins to stick, you can add a little water to deglaze it.

When the veg is softened and the spices are smelling fragrant, add the stock, coconut milk, tinned tomatoes, lemon juice, red lentils and grated root vegetables. Give it a good stir and cook for 20–30 minutes bubbling over a low heat. If you want a really intense flavour, leave on low overnight in a slow cooker.

While the dahl is cooking, pour the rice into a saucepan and cover with water three times the height of the rice in the pan and add a good couple of pinches of salt. Bring to the boil, then reduce to a simmer for 15–20 minutes.

Now make the flatbreads. Mix the flour, salt and oil together in a bowl. Slowly add water until you have a spongy dough that is easy to roll. Roll out on a well floured surface with a rolling pin, then heat a frying pan for 2 minutes over a high heat. Add the flatbreads, one at a time, and dry fry for 2 minutes on each side, or until the surface is slightly charred. Store under a tea towel until ready to serve.

In the last 10 minutes of cooking, add the cabbage or kale leaves to the dahl, if using. Check the dahl – depending on the stock, you may need to add more salt to taste.

Drain the rice and divide between 4 bowls. Ladle over the dahl and add some chopped coriander, pink pickled onions and shaved radishes on top. Serve with warm flatbreads.

Poached Pears with Buckwheat Crumble

For us, pears are one of the UK climate's most plentiful bounties. The UK is often associated with apples and cider, however, pears are the unsung heroes of British fruit. This recipe is one of Sadhbh's good friend Jenny's favourite desserts. With practice, we have learned that teaming sweet fruit and honey with creamy oat créme fraîche and the crunch of the buckwheat crumb makes a flavour and texture match made in gluten-free, vegan dessert heaven.

Prep: 15 minutes
Cook: 30–40 minutes

Serves 4

Ingredients

4 ripe pears, peeled, cored and halved
200 ml (generous ¾ cup) honey or maple syrup
1 orange, juiced and 3–4 strips of peel removed
1 star anise
1 cinnamon stick
A few slices of fresh ginger
Oat créme fraîche, to serve

For the buckwheat crumble:

40 g (1½ oz) non-dairy butter or coconut oil
80 g (generous ½ cup) buckwheat flour or cooked leftover buckwheat groats
25 g (scant ¼ cup) gluten-free oats
25 g (scant ¼ cup) any nuts or seeds, chopped
25 g (2 tbsp) muscovado sugar

Method

Preheat the oven to 200°C (180°C fan/400°F/Gas 6).

Place the pear halves in a lidded pan wide enough to lay them all next to each other, flat side down. Pour over the honey or maple syrup and the orange juice. Add the star anise, cinnamon stick, ginger slices and orange peel strips. Pour over a little water if the pear halves are not fully submerged.

Cover with the lid and set over a medium heat. Bring to the boil, then tlower the heat and simmer for 10–15 minutes. Turn the heat off but leave the lid on the pan to allow the pear halves to gently poach in the residual heat.

To make the buckwheat cumble, rub the butter or oil into the flour (or rub into whole leftover buckwheat groats for more of a crumb). Add the oats, nuts or seeds and the sugar, and mix together. Spread the mixture out onto a baking tray and bake for 15–20 minutes until just turning golden.

Sprinkle a few tablespoons of the crumble beside a couple of poached pear halves and add a generous dollop of oat créme fraîche to serve.

TIPS:

- If you want a more syrupy poaching liquor, remove the pears before reducing the spiced liquor further, boiling for another 5–8 minutes until it has a thicker consistency and can be poured over the créme fraîche.
- The pear peels can be reserved and chopped into muesli, added to a tagine or sweet stew, or eaten as a snack.

Caramelized Onion Tart

We have loved the challenge of making a great vegan tart, and although it's taken us a while, this version can really fool the most seasoned omnivore. The caramelized onions give a sweetness and the miso adds that salty umami that will have you reaching for another slice.

Prep: 30 minutes, plus chilling time
Cook: 60–90 minutes

Makes 1 x 23 cm (9 inch) tart or 4 x 11 cm (4¼ inch) tarts

Ingredients

3 tbsp oil of choice
4 onions, thinly sliced
3 tbsp balsamic vinegar
280 g (10 oz) silken tofu
250 ml (1 cup) soya cream
1 tsp miso, soy sauce or yeast extract (such as Marmite)
¼ tsp ground turmeric (optional,
Wholemeal flour, for dusting
1 quantity of Shortcrust Pastry (see page 84)
1 tbsp nutritional yeast
for a classic quiche colour)
Sea salt and ground black pepper
Vegan cheese (optional)
Seasonal edible flowers, to garnish (optional)
Heritage Tomato and Nasturtium Side Salad (see page 148), to serve

Method

Preheat the oven to 200°C (180°C fan/400°F/Gas 6).

Heat the oil in a large frying pan with a lid over a medium heat for a couple of minutes. Add the onions and balsamic vinegar, stir and turn the heat down low and put the lid on. Cook for 30–45 minutes until tender, stirring occasionally to prevent sticking.

In a blender, combine the silken tofu, soya cream, miso, turmeric, if using, and some salt and pepper. Blend until smooth.

Lightly dust a work surface and rolling pin with flour and roll out the pastry. Roll up the pastry on the rolling pin, and gently unroll it into the tart tin(s), pressing down into the corners. Trim the pastry, making sure you still leave some hanging over the edges. Transfer to the fridge to chill for 20 minutes.

Cut a section of greaseproof paper larger than the tin(s) and press into the pastry. Fill with baking beans (these can be substituted for rice or dried beans) and bake for 10–20 minutes until the base has dried.

Remove the greaseproof paper and baking beans from the tart case and bake for 5 more minutes. Remove the tart case from the oven and trim the edges of the pastry down to the tart tin.

Spread the onions evenly over the base of the tart and pour over the tofu mixture. Top with vegan cheese, if using. Bake in the oven for 30–40 minutes for a large tart, or 15–25 for small, until golden on top. Sprinkle over the edible flowers, if using, and serve with the salad.

Marrow Fritters with Roast Autumn Veg and Lentil Salad

Marrows are not usually seen as the sexiest vegetable. However, helping make veg sexy is what we do. Most marrows found at farmers' markets and some supermarkets are just overgrown courgettes – but are still tasty and nutritious, they just hold a little more water. As fritters, the marrow takes centre stage, showing its true potential.

Prep: 50 minutes
Cook: 40 minutes

Serves 4–6

Ingredients

1 medium marrow, or 2–3 courgettes (about 500 g/1 lb 2 oz), grated
¼ tsp sea salt
150 g (generous 1 cup) wholemeal flour
1 tsp baking powder
2 garlic cloves, crushed
2 spring onions, chopped (optional)
2 tbsp rapeseed oil
Ground black pepper

For the lentil salad:

500 g (1 lb 2 oz) autumn vegetables, chopped into bite-sized chunks (see Note overleaf)
½ tbsp oil of choice
200 g (generous 1 cup) puy lentils (or 2 x 400 g / 14 oz) cans green lentils, drained)
Bunch of parsley, leaves roughly chopped and stalks finely chopped, plus extra to serve

Method

Start by making the salad. Preheat the oven to 200°C (180°C fan/400°F/Gas 6).

Add your chosen vegetables to a roasting tray and toss with the oil so they are evenly coated. Roast for up to 45 minutes, depending on how browned and softened you like your veg. Check occasionally and shake the tray, to ensure consistent roasting.

When you've got the veg in the oven, start the lentils. In a large saucepan, cover the dry lentils with 5 cm (2 inches) of water. Bring to the boil then reduce the heat to a simmer. Puy lentils will take about 30 minutes to cook over a medium heat. Add ½ teaspoon salt to the lentils after the first 10 minutes of cooking, as salt can sometimes inhibit pulses cooking. To check when the lentils are cooked, spoon out a few, leave them to cool slightly, then try crushing them. If they are crushable and creamy soft, they are ready. If they are still quite hard and don't break open easily, they need longer.

With the veg roasting and lentils cooking, start your fritters. If the marrow is particularly overgrown, the skin can be tough and you may want to peel it and scoop out the seeds. Just-overgrown or large courgettes are preferential, and more flavourful.

Place the grated marrow in a colander over the sink. Add the salt and gently toss to combine. Leave it to sit for a few minutes, then use the back of a spoon or your hands to squeeze any excess juices out through the colander. Hands work better to get it all out.

Combine the marrow, flour, baking powder, garlic and spring onions, if using, in a large bowl and season with salt and pepper.

Heat the rapeseed oil in a large frying pan over a medium-high heat. Scoop 2 level tablespoons of batter for each fritter into the pan, flattening out with the spoon if needed. Cook for about 2 minutes, or until turning golden brown around edges and underneath. Flip over and cook on the other side for another 1–2 minutes.

[Ingredients and method continued overleaf]

For the vinaigrette:
3 tbsp smoked rapeseed oil
1 tbsp red wine vinegar
1 garlic clove, crushed
1 tsp wholegrain mustard

Place the cooked fritters on an oven tray and keep warm in the residual heat of the oven while you fry the remaining fritters.

To make the vinaigrette, place all the ingredients in a jar with a tight-fitting lid and shake vigorously. Set aside.

When the lentils are cooked, drain in a colander and toss with the roasted veg, the parsley and the vinaigrette. Leave to sit for a few minutes and toss again, scooping up all the dressing from the bottom of the bowl.

Serve on a platter with an extra sprinkle of parsley. Everyone can then help themselves to as much or as little of this filling salad as they want to go alongside their marrow fritters. Leftovers of this salad are great, as the dressing soaks into the lentils.

*** Note:** *we use butternut, kuri squash or pumpkin in this recipe as they're in season at the same time as marrow, and are actually from the same plant family, along with leeks or onion wedges. Later in the year we make this salad with Jerusalem artichokes, parsnips and carrots as they all roast in roughly the same time. Beetroot wedges are also great, but chop them slightly smaller, or start roasting them 10 minutes before, if roasting with other vegetables.*

'Waste Not' Einkorn Soda Bread

This recipe not only enables you to whip up a quick bread without long sourdough fermentation, it also allows you to use up any leftover dairy in the fridge. We've used buttermilk or soured milk, but, you can use kefir, yogurt or whey. The secret with soda bread is that the rise comes from the reaction of an acid with the bicarbonate of soda, so as long as you have acid in there, you will get some rise. If using regular milk or a plant-based milk, add a tablespoon of lemon juice or vinegar for every 250 ml (1 cup) milk and leave for 5–10 minutes to thicken. This will help the reaction along.

Prep: 15 minutes
Cook: 30–40 minutes

Makes 1 small round loaf

Ingredients

350 g (2½ cups) wholegrain einkorn flour, plus extra for dusting
1 tsp fine sea salt
1 tsp bicarbonate of soda
300 ml (generous 1¼ cups) buttermilk or soured milk

Method

Preheat the oven to 200°C (180°C fan/400°F/Gas 6). Line a baking tray with greaseproof paper.

Mix the einkorn flour, salt and bicarbonate of soda in a mixing bowl. Add the buttermilk, or equivalent, to the flour and stir until it all comes together into a dough. You might need to use your hands to bring it together.

Turn out the dough onto a floured surface and knead for a couple of minutes until it all comes together. Shape into a round and place on the lined baking tray. Use a sharp knife to score a deep cross on the top.

Bake in the oven for 30–40 minutes, or until there is a rise and some browning of the crust. When flipped over and tapped on the base it should sound hollow, indicating it's baked through. Leave to cool on a cooling rack before slicing.

Quick Creamy Mushroom Pasta

This recipe is adapted from Abi Aspen's partner Sean's staple comfort-food dish. One of the first dishes he made when he moved to the UK from Sweden, it's a quick and simple midweek pleasure that is so satisfying when you come home from a long day's work. The recipe lends itself well to alternative creams, so the vegan option is an easy swap.

Prep: 10 minutes
Cook: 20 minutes

Serves 2

Ingredients

3 tbsp oil
1 leek, thinly sliced with white and green parts separated
1 onion, diced
4 garlic cloves, thinly sliced
500 g (1 lb 2 oz) chestnut mushrooms
4 tbsp soy sauce
2 handfuls of greens (kale or chard work well), cut into strips
250 ml (1 cup) oat cream (or any other cream)
100 g (3½ oz) dried wholegrain pasta (or 150 g/5½ oz fresh pasta)
Sea salt and ground black pepper

Method

Heat the oil in a frying pan over a medium heat for a couple of minutes, then add the green part of the leek. Cook for 5 minutes until a little softened before adding the onion and white part of the leek. Cook until soft.

Stir in the garlic, mushrooms, 1 tablespoon of the soy sauce and a pinch of salt and pepper. Cover with a lid and let the mushrooms cook down over a low heat for about 10 minutes until soft. Once softened, add the rest of the soy sauce, the greens and the oat cream and stir.

While waiting for your mushroom mix to cook, bring a large saucepan of water to the boil and add 1 tablespoon salt. Add the pasta and cook for 10–12 minutes if dried or 3–5 minutes if fresh. Drain and serve the pasta in deep bowls with the oozy mushroom mix on top.

Beetroot and Goat's Cheese Tart

This tart is our famous *My Million Pound Menu* dish, where we explained to the judges about regenerative farming. We used heritage grains, vegetables and dairy to tell the story of holistic mixed farming and how plants and animals interact in a sustainable food system. Aside from the environmental inclination, this tart is both delicious and easy to make.

Prep: 30 minutes, plus chilling time
Cook: 60–70 minutes

Makes 1 x deep 23 cm (9 inch) tart

Ingredients

400 g (14 oz) beetroot, cut into wedges
3 tbsp oil of choice
Small bunch of thyme, leaves picked
1 tbsp wholemeal flour, for dusting
1 quantity of Shortcrust Pastry (see page 84), chilled
5 medium eggs
300 ml (1¼ cups) double cream
150 g (5½ oz) goat's cheese
Sea salt and ground black pepper

Method

Preheat the oven to 200°C (180°C fan/400°F/Gas 6). In a roasting tin, toss the beetroot in the oil with the thyme and a pinch of salt and pepper. Roast in the oven for 30 minutes until tender and slightly browned.

Lightly dust a work surface and rolling pin with flour and roll out the pastry. Roll up the pastry on the rolling pin and gently unroll it into a 23 cm (9 inch) tart tin, pressing down the into the corners. Trim the pastry, making sure you still leave some hanging over the edges. Transfer to the fridge for to chill 20 minutes.

Cut a section of greaseproof paper larger than the tin and press it into the tart. Fill the tart with baking beans (can be substituted for rice or dried beans) and bake for 10–20 minutes until the base has dried.

Whisk the eggs and cream together in a bowl and add a hearty pinch of salt and pepper.

Remove the greaseproof paper and baking beans from the tart case and lightly brush some of the egg mixture over the inside of the tart. Bake for 5 more minutes until the egg mixture has set the pastry. Remove the tart case from the oven and trim the edges of the pastry down to the tart tin.

Arrange the roasted beetroot evenly in the tart and pour the rest of the egg mixture over the top, not quite up to the rim of the pastry. Crumble the goat's cheese evenly on top. Bake for 30–40 minutes until golden and the egg mixture doesn't wobble when shaken anymore. Serve warm or at room temperature.

Buckwheat Waffles with Rhubarb, Sheep's Labneh and Pumpkin Seed Brittle

Sadhbh first had delicious, homemade waffles in the mountains of Vermont, with foraged wild blueberries. She was hooked, and so the hunt began. Within a few days of arriving home she'd found a secondhand waffle iron on Gumtree for a fiver. We adapted this recipe to use buckwheat flour, making it an ideal, decadent gluten-free breakfast for a wild, private birthday weekend we catered in a huge, English country house. That was definitely a classic work-hard-play-hard weekend for The Sustainable Food Story, with all of us dancing as much as cooking. We often opt for sheep's or goat's yogurt when making labneh, to reduce our cow's milk dairy consumption, and it's sharpness complements this sweet breakfast.

Cook: 30 minutes
Prep: 40–50 minutes, with advance prep the day before

Makes: 8 waffles

Ingredients

7 g (2 heaped tsp) fast-action yeast (or replace with 1 tbsp Sourdough Starter, see page 151)
120 ml (½ cup) lukewarm water
8 tbsp butter or vegan spread
470 ml (scant 2 cups) milk of your choice
1 tsp sea salt
1 tsp coconut sugar
240 g (1⅔ cups) buckwheat flour
2 medium eggs
¼ tsp bicarbonate of soda

Method

The day before:

First, prepare the waffle batter. Mix the yeast with the warm water in a large bowl and wait for it to dissolve. Melt the butter in a saucepan, add the milk, and warm gently, then add all this to the yeast mixture.

In a second mixing bowl, whisk the salt and sugar into the flour. Add this to the liquid and whisk until smooth. Cover the bowl and leave it to stand overnight at room temperature (if making with sourdough starter this should also get active and bubbly overnight).

To make the sheep's labneh, pour the yogurt into a clean muslin cloth and gather it into a ball, squeezing out the moisture as you tighten the thin cloth around it. Tie the cloth with an elastic band, and then sit it in a sieve over a sturdy bowl and put a weight on top, such as a ramekin. Transfer it to the fridge and leave overnight. The weight will help drain more whey out of the yogurt.

To make the rhubarb, mix the maple syrup and coconut sugar with 100 ml (scant ½ cup) water in a saucepan and bring to the boil. Add the rhubarb to the boiling syrup, give it a good stir, then cover with a lid and take the pan off the heat. Leave the rhubarb in the sweet syrup as it cools, so it gently poaches.

To make the pumpkin seed brittle, toast the pumpkin seeds for a few minutes in a frying pan over a medium heat. Spread the toasted seeds out evenly, but densely packed, on a sheet of greaseproof paper on a baking tray.

In a small heavy-based pan, melt the sugar, butter and salt with 3-4 tablespoons water. Heat it up gently, making sure all the sugar is dissolved but none is catching around the sides of the pan. Bring the

[Ingredients and method continued overleaf]

For the sheep's labneh:
1 pot sheep's yogurt (we use Woodland's organic)

For the rhubarb:
150 ml (⅔ cup) maple syrup
3 tbsp coconut sugar
4–6 long stems of pink rhubarb, cut into 3 cm (1¼ inch) batons

For the pumpkin seed brittle:
100 g (scant ⅔ cup) pumpkin seeds
80 g (scant ½ cup) caster sugar
40 g (1½ oz) butter
Pinch of sea salt flakes

caramel up to a boil for a few minutes, until it turns golden brown, swirling to stir it.

Tip the caramel evenly over the toasted seeds on the tray. It will start setting almost instantly. Leave to cool fully before cracking into shards. This can be done by putting the point of a knife into the middle of the caramel sheet and tapping from the handle end, to get a crack down the middle. Be careful not to bend the tip of your knife – use a thick-tipped knife, or even a clean hammer.

The next morning:
Unwrap the yogurt and you'll be left with sheep's labneh that can be scraped from the muslin, ready to dollop on the waffles.

When you are ready to make waffles, beat the eggs and bicarbonate of soda into the batter mixture.

Ladle a good spoonful of batter into a very hot waffle iron and cook until the waffle is crisp. Check after about 3 minutes to see if it's turning golden. Continue until you have made 8 waffles, keeping them warm as you cook in batches.

Serve the crisp waffles topped with dollops of labneh and poached rhubarb and sprinkled with shards of pumpkin seed brittle.

TIPS:

- Cloe, Sadhbh's domestic goddess mum, recently introduced us to her hack for making pale forced rhubarb a more brilliant pink – add hibiscus flowers. They even have a similar tang to rhubarb.
- This waffle batter will keep for days covered in the fridge, and can of course be used like any batter – to make pancakes, toad in the hole, or leftover dollops can be used to bind fritters, or as a tempura for frying veg.
- The whey can be collected and used in the Whey Cheese Sauce (see page 130) or to supplement the buttermilk or yogurt quantity in soda bread (see page 113) or any creamy sauce really.
- You can keep the syrup from the poached rhubarb to reduce and pour over as a syrup, or reserve to add to a cocktail (pink rhubarb G&T is fab).

Zero-Waste Orange Cake

This recipe is what we like to call a 'gateway vegan cake' – a dipping of your toe into dairy alternative cooking waters. It is adapted from Sadhbh's orange marmalade cake, however, instead of marmalade we purée the whole orange. The trade-off for a little more prep time means you don't waste any of your orange, and the purée gives a super-moist, moreish cake. The nuts add to that moistness, and the type of nuts you use is quite flexible, depending on your location and availability. We like to use cobnuts in the UK, which Abi Aspen spends blustery September days picking in Cornwall every year with her mum Averil, for a sustainability-focused option.

Prep: 20 minutes
Cook: 2½ hours

Makes 1 x 28 cm (11 inch) cake

Ingredients

- 1 medium unwaxed orange
- 175 g (6 oz) vegan block or butter, softened
- 75 g (generous 1/3 cup) muscovado sugar
- 100 ml (scant ½ cup) maple syrup
- 1 flax egg (see note)
- 1 medium egg
- Finely grated zest of 1 orange or lemon
- 75 g (generous ½ cup) wholemeal flour
- 100 g (¾ cup) cobnuts or hazelnuts, finely blitzed
- 1 tsp baking powder
- 100–200 g (¾–1½ cups) seasonal fruit, sliced

Method

In a small saucepan over a medium heat, cover the orange with water, pop on the lid and boil for 2 hours, making sure to keep the water topped up. We usually do a few oranges together to make it worth the boil. Transfer the orange to a blender, making sure to remove any pips, and blitz until smooth.

Preheat the oven to 175° C/155° C fan/350° F/Gas 4. Combine the vegan block or butter with the sugar and maple syrup in a mixing bowl and beat until fluffy and creamy, then add the flax egg, egg, orange zest and the puréed orange. Mix the flour with the cobnuts/hazelnuts and baking powder then fold into the cake mix.

Line a 28 cm (11 inch) cake tin with greaseproof paper, and fill with the cake mixture. Top with the slices of seasonal fruit, gently pressing into the top of the cake. Bake for 20–30 minutes, until slightly browned and a knife comes out of the centre clean. The cake will be quite flat, but will taste all the better for it!

* *NOTE: to make a flax egg, mix 1 tablespoon of ground flax with 3 tablespoons of water then let the 'egg' sit for 10 mins to coagulate.*

Speltotto for Every Season

We've worked with some grain diversity loving chefs over the years. Johanna (Sadhbh's Danish earth mother, chef colleague at the Skip Garden) and Maisie (Abi Aspen's crumpet-queen mentor at cafe Good and Proper Tea – now running D'oh Life) are two such inspiring grainiacs. Spelt, pearl barley or other whole grains make great risottos. Try roast beetroot and beet leaf, pea, lemon zest and mint as seasonal variations here.

Prep: 10 minutes
Cook: 25 minutes

Serves 4

Ingredients

850 g (1 lb 14 oz) butternut or kuri squash, deseeded and diced
4–6 tbsp oil
1 onion, diced
3–4 garlic cloves, crushed
About 30 sage leaves – half chopped, half left whole
200 g (1 cup) spelt, rinsed
800 ml–1 litre (3–4 cups) vegetable stock (option to substitute 100 ml/scant ½ cup with white wine)
100–150 g (3½–5½ oz) goat's cheese (optional)
Sea salt and ground black pepper

Method

Preheat the oven to 200°C/180°C fan/400°F/Gas 6.

In a large roasting tin, toss together the squash, 2–3 tablespoons of the oil and some salt and pepper until well coated. Roast for 20–25 minutes until the skins of the squash are soft and easy to eat.

Meanwhile, heat another 1–2 tablespoons of the oil in a large saucepan over a medium heat and sauté the onion for 5 minutes. Add the garlic and chopped sage leaves.

Add the spelt to the onions, sautéing for a few minutes to brown the grains. Pour in the white wine, if using, and cook off for a few minutes, before adding 300 ml (generous 1¼ cups) of the stock. Bring to the boil, stir, then turn down the heat and simmer for 10–15 minutes, stirring occasionally. Pearled spelt will be ready faster than an unpearled variety, as it has less fibre.

Once most of the stock has been absorbed, add more stock bit by bit, until the grains are cooked to your preferable texture, much like arborio rice for risotto.

Once the grains are nearly cooked, add the roasted squash. If choosing other flavour combinations, add at this point instead of the squash. Add half the cheese at this stage, if using.

In a medium frying pan, heat the remaining oil. Once hot, drop in the remaining whole sage leaves for a minute or less to crisp them up. Take them out before they turn dark brown.

To serve, spoon the speltotto into wide bowls or plates. Garnish with a little goat's cheese, if using, and top with the crispy fried sage.

TIPS:

- This recipe lends itself well to making croquettes. Roll golf-ball sized balls of leftover speltotto in breadcrumbs and deep-fry for a few minutes until golden brown.
- If using other ingredient combinations, substitute sage for another herb – try thyme with mushrooms, parsley for beetroot, lemon zest if using greens or mint with peas.

Lentil Moussaka

Prep: 30 minutes
Cook: 40–50 minutes

Serves 4–6

Ingredients

120g (¾ cup) red, brown or puy lentils
2 large aubergines, thinly sliced lengthways
4 tbsp oil
3 shallots, finely chopped
1 red pepper, chopped
2 garlic cloves, finely chopped
3 tbsp tomato purée
400 g (14 oz) tin chopped tomatoes
1 tsp ground cinnamon (or 1 cinnamon stick)
1 tsp mixed dried herbs (or a few leaves of fresh herbs such as oregano, thyme or parsley)
Sea salt and ground black pepper
Your favourite cooked greens or a dressed salad, to serve

For the white sauce:

200g (7 oz) ricotta
2 medium eggs or 1 whole egg and 1 egg yolk (if already using an egg white for binding granola, see page 144)
Pinch of freshly grated nutmeg
50 g (2 oz) freshly grated hard cheese, like kefalotyri, parmesan or pecorino

Sadhbh's dad, Norma, used to make a mean moussaka when she was growing up. Despite the fact that she didn't enjoy meat very much, the flavours all worked so well together that it became a favourite dish. We don't miss the meat in this version, with a veggie or vegan white sauce topping the umami layers. The quintessential elements of moussaka are all there: succulent layers of griddled aubergine enveloped in oil, textured tomatoey layers with sweet cinnamon and mediterranean herbs, and that hint of nutmeg in the creamy crisped topping.

Method

Preheat the oven to 200°C (180°C fan/400°F/Gas 6).

Make the filling first. Cook the lentils in a small saucepan, covered in at least three times their volume in boiling water until just cooked – 5–10 minutes for red lentils, 15–20 minutes for brown lentils or 20–25 minutes for puy lentils. Strain and set aside for adding to the sauce later.

Use a pastry brush to generously coat the aubergine slices with about 3 tablespoons of the oil. They suck it up like a sponge, but need it to cook. Season the slices with salt and pepper.

Sear the aubergine slices in a large frying pan over a medium-high heat for a few minutes on each side, until they are turning golden and soften. Do this in batches if necessary to avoid crowding the pan. Alternatively, griddle the slices on a griddle pan until they have dark brown char lines.

Heat the remaining oil in a medium saucepan. Add the shallots and fry for 3–4 minutes, before adding the red pepper and garlic and frying for another couple of minutes. Then add tomato purée, tinned tomatoes, cinnamon and herbs. Bring to the boil. Add the lentils and season with salt and pepper. Simmer for 5–10 minutes.

While the lentil tomato sauce reduces, make the white sauce by mixing the ricotta, eggs, nutmeg, some salt and pepper and half the grated cheese together in a bowl.

To assemble, layer half the cooked aubergine slices in the base of a shallow ovenproof dish. Add a layer of lentil tomato sauce. Add the next layer of aubergine slices and then top with the white sauce and sprinkle with the remaining cheese.

Bake in the oven, uncovered, for 20–30 minutes, until it is golden and bubbling on top. Serve hot with a side of greens or a fresh, dressed salad.

Basil Panna Cotta with Charred Balsamic Strawberries

Strawberries and goat's dairy are good pals – the sweetness and tangy dairy richness really compliment each other. We often use goat's or sheep's dairy to vary our demand for cow's dairy. This dessert works well with dairy-free milks too. We originally made this recipe with foraged sheep's sorrel, but it's not as easy to come by, so we've given quantities for infusing with basil, an excellent alternative flavour.

Prep: 10 minutes
Cook and cool: 30 minutes, plus chilling time

Serves 4

Ingredients

350 ml (scant 1½ cups) goat's milk (or dairy-free milk)
½ vanilla pod, split in half
Grated zest of ¼ lemon
200 g (7 oz) sheep's sorrel (or 3 sprigs of basil)
70 g (⅓ cup) caster sugar
1 sachet vegan gelatin (or carrageen/Irish moss)
200 ml (generous ¾ cup) goat's, oat or coconut cream (we have had the best results with oat cream)
Vegetable oil, for greasing

For the charred strawberries:

1 tbsp goat's butter, vegan block or coconut oil
1 x punnet (250 g/1¼ cups) strawberries, halved
25 g brown sugar
2 tbsp balsamic vinegar

Method

In a medium saucepan, combine the milk, vanilla pod, lemon zest, sheep's sorrel or basil and sugar. If you're using carrageen/Irish moss as the vegetarian setting agent, it needs to be soaked in water for 15 minutes then added to the milk at this stage. Bring the mixture to a simmer, but don't boil. Turn off the heat and allow to infuse for 15 minutes. Strain the infused milk through a sieve, into a measuring jug, retaining the vanilla seeds, but leaving the lemon zest, herbs and vanilla pod (and carrageen, if using) behind in the sieve. If using a sachet of vegan gelatin, follow the instructions on the package and add to the milk mixture. Mix in your cream of choice. Lightly oil four 175 ml (6 fl oz) ramekins or similarly sized pudding basins. Pour the mixture into these moulds.

Transfer to the fridge and chill for at least 30 mins, or until set.

Meanwhile, make the charred strawberries. Melt the butter or coconut oil in a pan over a high heat. Tip in the strawberries and sprinkle with the brown sugar, giving it a good mix to evenly coat. Leave to fry, caramelizing the sliced sides of the strawberries for a couple of minutes. This is a perfect way to use up slightly soft strawberries. Finally, add the balsamic vinegar and wait for the liquid to thicken into a caramel.

Veggie setting agents don't set exactly like gelatine, so expect that it may be quite a different texture to traditional Italian panna cotta. Unmould the puddings by sliding a knife around the edge of the ramekins, turning them upside down over a serving plate and gently easing them out, pulling the deserts down to release them.

Serve the panna cottas with the syrupy charred strawberries poured over and around. Garnish with any remaining sheep's sorrel or basil leaves.

TIP:

- Sprinkle over granola, leftover buckwheat crumble from page 107 or broken biscuits for a bit of crunch, if liked.

Chicken Liver Pâté

When we do end up in a supermarket, we often notice that chicken livers are discounted due to apparent unpopularity. As we came to research offal, we realized that if we are going to eat meat, we need to eat 'nose-to-tail'. Offal meat is generally the most nutritious part of an animal, with liver and kidneys having some of the greatest nutrient density. Eating offal is also a great way of adding meat to your diet without contributing to the overall consumer demand.

Prep: 20 minutes
Cook: 15 minutes

Serves 4–6

Ingredients

200 g (7 oz) butter
1 small onion or shallot, diced
1 tart apple, peeled, cored and diced
2–3 garlic cloves, sliced
400 g (14 oz) chicken livers (we use organic free-range), trimmed of any sinews
3 thyme sprigs, leaves picked (or ½ tsp dried thyme)
3 tbsp port or Madeira or 2 tbsp brandy
Pinch of ground allspice, if you have it
Sea salt and ground black pepper

Method

Heat 1 tablespoon of the butter in a non-stick frying pan over a medium heat. Add the onion and apple, cooking for about 5 minutes until soft and lightly browned. Add the garlic and fry for a further 2–3 minutes. Tip the onion mixture into a food processor or blender.

Return the frying pan to a medium heat and add another tablespoon of the butter. Once the butter is melted and foamy, add the chicken livers and fry for 4–5 minutes, turning after 2 minutes.

Add the thyme, alcohol, allspice, if using, and some salt and pepper and cook for another minute or so until the livers are just done. They should have turned from red-pink to brown, but should still be soft.

Tip this mixture into the food processor or blender with the onions. Reserve 2 tablespoons of the butter but add the rest to the processor with the warm ingredients. Blend all the ingredients to a smooth paste. Taste and adjust seasoning if necessary.

If you're planning to use it immediately, leave it to cool slightly, but it will also cool once spread. Alternatively, store in a sterilized clip-top jar (see page 55).

Melt the reserved butter to pour over the smoothed final amount to store, as this will seal it and delay spoilage.

** **NOTE:** the paté will keep for up to 3 days in the fridge. We like to freeze the paté in ice cube trays to consume one nutrient-boosting portion at a time, when we haven't eaten red meat in a while. Keep in the freezer for up to 3 months; defrost in the fridge overnight, or on the lowest 'defrost' microwave setting for 30 seconds at a time until defrosted throughout. It's delicious topped with bread and butter.*

Roasted Squash with Winter Greens, Black Pudding and Whey Cheese Sauce

We made this dish for the lovely women's group at Farmdrop, for a late Christmas gathering. As a food supplier, they had a great kitchen, loads of knowledge of great ingredients and were really enthusiastic to hear the stories behind why we chose the dishes we did.

We've made our own blood pudding (or black pudding, as it's called in the UK, due to the dark colour the blood turns when cooked) in the past – ironically for a cooking demo at Shambala Festival, the only vegetarian major festival in the UK at the time. We delved into loads of research around offal, byproducts, abattoirs and traditional blood pudding makers, and blood pudding around the world, like *boudin noir*, *mutura*, *morcilla* and *bloedworst*. It's hard to get hold of fresh blood to make the real deal at home, unless you live near a farm. In the UK, Fruit Pig are our favourite supplier of good-quality blood pudding made with fresh blood, using free-range fat and British grains.

The whey cheese sauce in this recipe was inspired by the duo Josh and Mike at Blanch and Shock, who use their Aladdin's cave food studio to turn waste into delicacies. It adds a delicious cheesy covering for the squash and blood pudding.

Prep: 1 hour
Cook: 2 hours

Serves 4

Ingredients

1 kuri squash (or butternut squash if that's what's available), deseeded and cut into 2 cm (1 inch) crescent wedges
2 tbsp oil
3–4 rosemary sprigs
Sea salt and ground black pepper

Method

Preheat the oven to 200°C (180°C fan/400°F/Gas 6).

In a large roasting tray, toss together the squash wedges with the oil, some salt and pepper and the rosemary so they are evenly coated then arrange so that the wedges are standing up. Roast for 30–40 minutes until crisping at the edges, soft enough that a knife goes through easily and the skin is soft.

To make the winter greens, heat the oil in a pan over a medium heat and add the sliced greens and garlic. Cook for 5–10 minutes until softening and releasing juices, but not too limp. Season with salt and pepper. Dress with the lemon juice just before serving.

Next prepare the blood pudding. Heat the oil in a frying pan over a medium heat and fry the blood pudding on both sides for 3–4 minutes until crispy.

[Ingredients and method continued overleaf]

For the winter greens:
2 tbsp oil
1 winter cabbage, such as a pointed cabbage or small savoy cabbage, cut into roughly 1 cm (½ inch) wide slices
3 garlic cloves
Juice of ½ lemon

For the blood pudding:
400 g (14 oz) good-quality blood pudding, sliced into 1.5 cm (¾ inch) thick rounds
2 tsp oil

For whey cheese sauce:
1 litre (4 cups) whey (you can ask a cheesemonger for whey if you are not making it yourself)
Skins of 10 garlic cloves
100–200 g (3½–7 oz) veg trimmings (optional)

In a medium saucepan, combine all of the ingredients for the whey cheese sauce and bring it to the boil. Keep at a boil until the liquid has reduced to one-quarter of the amount you started with. Taste and season, then strain the sauce into a jug.

Divide the greens between bowls and top with the squash wedges and black pudding. Serve the whey sauce in a jug, allowing diners to help themselves to the tangy, creamy accompaniment.

Goat or Wild Venison Casserole

Wild venison or goat are our go-to red meat options if we feel we need a good dose of haem iron. The best thing about a casserole is that there are so many other flavours to mellow out sometimes strong-tasting game, and it cooks in the oven for a few hours so you can get on with other things. A more energy-efficient way to cook a meat stew is in a slow cooker. If you're using an oven, consider making a big batch of this casserole for freezing, or bake something else alongside it on another shelf.

Prep: 30–60 minutes
Cook: 3–3½ hours

Serves 4

Ingredients

800 g (1¾ lb) diced wild venison or goat, a mix of neck, breast, shoulder and leg
2–3 tbsp oil
3 onions, diced
2 celery sticks (or 1 similarly sized chunk of celeriac, diced)
3 carrots, diced
5 garlic cloves, crushed
8 chestnut mushrooms, sliced
1–2 tbsp wholemeal flour
1 rosemary sprig (or 1 tsp dried rosemary)
2 thyme sprigs (or 1 tsp dried thyme)
6 juniper berries, lightly crushed
3 bay leaves
200 ml (generous ¾ cup) red wine
300 ml (generous 1½ cups) bone broth or vegetable stock
2 tbsp tomato purée
Sea salt and ground black pepper
Bread, dumplings or grains, to serve

Method

Preheat the oven to 200°C (180°C fan/400°F/Gas 6).

Season the meat with a generous amount of salt and pepper. This can even be done the night before, to help tenderize the meat and marinate, but isn't essential.

In an ovenproof pan with a lid or casserole dish, heat half the oil over a medium heat and add the diced onions. Gently sauté the onions for 10 minutes before adding the celery, carrots, garlic and mushrooms. Sauté together until the rest of the veg is browning in parts and the carrots are starting to soften, then remove the vegetables from the pan and set aside.

Coat the seasoned meat in the flour. Wipe out the casserole dish and add the remaining oil, turn the heat to high and add the coated meat. Sear until the meat is browning on all sides – you may need to do this in batches. Add the herbs and juniper berries once the meat is browned.

Pour in the wine, followed by the stock and tomato purée, stirring it all together and deglazing the pan. The flour coating on the meat will help to create a thicker gravy. Add the veg mix back to the pan, making sure all ingredients are submerged in the liquid. Taste to make sure the seasoning is right and add more stock if needed, or water if salt levels are already adequate.

Bring the mixture to a simmer. Put the lid on and cook in the oven for 2½–3 hours, occasionally stirring and checking the moisture levels, and topping up with a little water or stock if necessary. The cooking time needed to reach a melt-in-the-mouth texture for the meat will be dependent on a range of factors, including age, breed and even sex of the deer or goat. A good casserole should have tender meat and the stewing sauce around it should be dark, thick and aromatic.

Serve with homemade bread, dumplings or grains of your choice, or top with pastry and bake for 20–30 minutes to turn it into a delicious pie

Shellfish Spelt Spaghetti

Shellfish may be one of the best sources of haem iron a non-red-meat eater can get. Mussels have been calculated to have a far lower carbon footprint from farming than many other sources of animal proteins, the formation of their shells even sequesters carbon. Growing up on the Atlantic coast, we occasionally picked fresh mussels or periwinkles for dinner. This recipe is adapted from the classic Italian dish of *spaghetti alle vongole*.

Prep: 20 minutes, plus soaking
Cook: 10–15 minutes

Serves 2

Ingredients

500 g (1 lb 2 oz) mussels and/or clams (MSC or highest sustainability assurance available)
2 tbsp sea salt
150 g (5½ oz) wholemeal spelt spaghetti (or 200 g/7 oz) fresh, wholewheat linguine)
3 tbsp extra virgin olive oil
3–4 garlic cloves, sliced
100 g (3½ oz) cherry or vine tomatoes, quartered
½ mild red chilli, chopped, or ½ tsp dried chilli flakes (optional)
100 ml (scant ½ cup) white wine (optional)
Small bunch of parsley, stalks finely chopped and leaves roughly chopped
Juice of 1/2 lemon, plus extra wedges to serve
Ground black pepper
Dried chilli flakes, to serve

Method

Clean and sort through the mussels and/or clams, removing any with broken shells or those that don't close when they're tapped. Place them all in a deep bowl of cold water with 1 tablespoon of salt for 30 minutes. Any that float to the top should be discarded. If they're freshly hand-picked from rocks, you'll need to de-beard them, too, which entails pulling off the fibrous hairs that they produce to attach to rocks.

Bring a pan of water to the boil, and once boiling add the remaining tablespoon salt. Add the pasta when the water is bubbling and cook until the pasta is al dente – this is 6–8 minutes for wholemeal spelt pasta, but will be only a few minutes for fresh pasta. The salt stops the pasta from sticking together, and contrary to popular belief, oil isn't needed when boiling pasta.

In a large, wide-based pan with a fitted lid, heat the olive oil, then add the garlic and fry for a minute. Add the tomatoes and chilli, if using, followed by the mussels and/or clams. Add a little water or the white wine at this point, if using, and give it a good shake to stop the tomatoes sticking.

Put the lid on and steam the shellfish for 3–4 minutes until they open. If any of the shells remain closed after cooking, throw them away. This is an indicator they aren't edible.

Drain the cooked spaghetti and gently stir it into the pan of lightly cooked tomatoes, garlic and shellfish. Sprinkle over the parsley. Squeeze the lemon juice. Serve immediately with the lemon wedges and a sprinkle of dried chilli flakes, if you want an extra kick.

Dog Nip

Abi Aspen made this for her puppy during training as she didn't want to use the suggested boiled chicken, choosing instead to create something offal-based, with ingredients that may otherwise be wasted. These treats are amazing for dog training and so easy to make.

Prep: 10 minutes
Cook: 30 minutes

Makes 1 x 34 x 27cm (13½ x 10½ inch) tray

Ingredients

- 400 g (14 oz) chicken livers (or other organ meat such as kidneys, hearts and spleens)
- 200 g (1½ cups) steel-cut oats (or any type of oats)
- 200 g (1½ cups) wholemeal flour
- 3 tbsp oil
- 1 medium egg

Method

Preheat the oven to 200°C (180°C fan/400°F/Gas 6). Line a baking tray with non-stick greaseproof paper.

Place all the ingredients in a food processor or blender and blend until semi-smooth. Alternatively, mince the chicken livers with a knife and mix in a bowl with all the other ingredients.

Spread the blended mixture in an even layer on the lined tray, to about 1 cm (½ inch) thick. Bake in the oven for 30 minutes. Leave to cool, then cut into small strips, which are easy to carry on walks for smaller pieces to be broken off.

Store in the fridge in an airtight container for up to a week. Any excess treats can be stored in the freezer for up to 6 months.

Tomato Sauce Base

This sauce is perfect with meatballs (see page 98), can be adapted for moussaka (see page 125), veggie or vegan lasagnes, lentil bolognese or venison ragu, or used as a simple tomato sauce for pasta bakes.

Prep: 5 minutes
Cook: 30 minutes

Serves 4–5

Ingredients

1 tbsp oil
1 large onion, diced
3 garlic cloves, crushed
1 tbsp sundried tomato pureé (optional)
2 x 400 g (14 oz) tins tomatoes, chopped or plum
2 tsp mixed dried Italian herbs
Sea salt and ground black pepper

Method

Heat the oil in a large saucepan. Add the onion and cook over a low-medium heat for about 10 minutes until soft, before adding the garlic and sundried tomato purée, if using. Cook for 2 minutes before adding the tinned tomatoes and herbs. The tomatoes can be broken up in the pan with a wooden spoon, or chopped using scissors (either in the pan or while still in the tin). Season with salt and pepper to taste, and simmer for 20–30 minutes.

This versatile sauce can be stored in a large, sterilized jar (see page 55) for up to a week in the fridge, or can be frozen for 3–6 months.

TIPS:

- To add more flavour, nutrition and to use up odds and ends of veg, we like to sauté a sofrito – a mix of a chopped celery stick or two, or celery core and a diced carrot or two. We also often add in a handful of chopped mushrooms, or even diced celeriac, leek and fennel, at the stage when gently frying the onions.
- To get extra veg into the likes of a kids' pasta bake or lasagne, we often blend this veg into the tomato sauce. This has been key to getting kids in cookery classes who claim they 'don't like vegetables' to eat them, in disguise. They often enjoy being in on the secret of the hidden veg, and enjoy getting others to inconspicuously eat them, too.

Vegan White Sauce

This works as a perfect substitute to the traditional dairy-based béchamel, and is perfect for adapting pies, lasagnes and moussakas to be more plant based.

Prep: 5 minutes
Cook: 5 minutes

Makes enough for 4–5 people for pie, lasagne or moussaka.

Ingredients

400–500 ml (1½–2 cups) soya or oat milk
2 bay leaves
5 peppercorns
Veg trimmings (such as tops, tails and skins of garlic, onions, fennel, leek, celery and carrot)
50 g (⅓ cup) plain flour
50 g (2 oz) fat – either vegan block or butter or oil

Method

Place the milk in a saucepan with the bay leaves, peppercorns and veg trimmings. Warm over a medium heat until just before boiling point, then reduce to a simmer for a few minutes. Turn the heat off and leave it to infuse.

Next make a roux by combining the flour and fat in a saucepan over a medium heat and cooking for 2–3 minutes until it makes a somewhat dry, beige paste. This step cooks the raw flavour out of the starch in the flour, and the fat helps to disperse the flour, allowing even thickening throughout the sauce.

Very slowly, add the warmed milk in a steady stream, whisking continuously to avoid lumps forming. If it ends up lumpy, it can be blended, so is always resurrectable. If it still seems thin, heat for a few more minutes until it thickens and coats the back of a wooden spoon.

TIPS:

- To prevent the white sauce forming a skin and getting lumps from that skin, cover with a cartouche while cooking, which can be made from any leftover greaseproof paper from bread or cookie baking. A cartouche is a round of greaseproof paper that has been scrunched up and dampened, and is placed on top of a white sauce or custard as a great alternative to cling film.

Pea and Mint Dip

Prep: 10 minutes
Cook: 5 minutes

Serves 10

Ingredients

400 g (3⅓ cups) frozen or fresh peas or broad beans
50 ml (scant ¼ cup) extra virgin olive oil (or your favourite dressing oil)
Juice of 1 lemon
Small handful of mint leaves
Pinch of dried chilli flakes (or ½ small red chilli, chopped)
Rye Levain Crackers (see page 142), to serve

This is an ideal recipe for peas or broad beans – it couldn't be easier. Baby broad beans can even be eaten in their pod, just like peas. The ever-innovative Nozzer (Norman – he of the potato moussaka, page 125) has even been known to make broad-bean-pod wine on occasion!

Method

Bring a saucepan of water to the boil and cook the peas or broad beans for 4–5 minutes. Strain and then whizz the peas/broad beans, oil, lemon juice, mint and chilli together in a food processor or blender or smash together in a large mortar and pestle. Leave as chunky as you like, or process to a smooth purée or pâté texture.

Use as a toast topping, as a canapé or in lunchboxes for dipping crudite, croquettes or crackers.

Pink Pickled Onions

Prep: 5 minutes
Cook: 10 minutes

Makes 1 x 340g (12 oz) jar

Ingredients

100 ml (⅓ cup) hot water
2 tsp golden caster sugar
1 tsp sea salt
100 ml (⅓ cup) cider vinegar
1 medium red onion, thinly sliced in half rounds
Ground black pepper

Homesteader and former Skip Garden chef Erin, now The Edible Flower, taught us these quick pickles. They are fab in dahl (see page 104), or for jazzing up the top of grain salads.

Method

Pour the hot water into a heatproof jug. Add the sugar and salt and stir. Once all their crystals have dissolved, add vinegar and a couple of grinds of black pepper.

Place the onions in a bowl and then pour over the brine. Press the onion slices down so they are completely submerged.

Leave the onions to quickly pickle for at least 10 mins. They can be used immediately, or can be prepared in advance.

Transfer any leftover onions to an airtight container, ensuring they're covered in the pickling liquid and store in the fridge. They will keep for several weeks.

Rye Levain Crackers

Often when you feed your sourdough starter, you are advised to discard some of the current starter to add more flour and water. To us, that 'waste' is tasty food! A fast-growing food trend we noticed with the lockdown sourdough baking frenzy, was making 'trash' baked goods – using surplus sourdough starter. We like to make levain crackers out of them.

Our favourite flour to feed our starter with is rye, as it has a strong flavour and creates an active culture. This recipe was inspired by working in the bakery with Dan Barber's team at Blue Hill, in Upstate New York, where nothing goes to waste. It's hard to specify exactly how much flour and water you will need as many people feed their sourdough starter by eye rather than weight, but the quantities given here assume a 100 per cent hydration sourdough starter (one part water for every one part flour). We've found that a silicone mat is easier for baking these than greaseproof paper.

Prep: 15 minutes
Cook: 10 minutes

Makes 20 large crackers

Ingredients

- 200 g (7 oz) rye Sourdough Starter discard (see page 151)
- 50 g (scant ½ cup) wholemeal flour, plus extra as needed
- Hearty pinch of fine sea salt
- Pinch of flaky sea salt, for sprinkling on top

Method

Preheat the oven to 200°C (180°C fan/400°F/Gas 6). Line a baking tray with a silicone mat.

Mix the starter discard and flour in a mixing bowl with the fine sea salt. The mixture should slowly drip off a spoon if lifted up. If it is too runny, add a little more flour; if too thick, add a little water.

Spread the mixture thinly (less than 1 mm, as thin as you can) evenly over the mat. Sprinkle the surface with the flaky sea salt. Bake for 5–10 minutes, or until lightly browned.

Leave to cool on the tray, then break the crackers into shards for serving. These work well as a snack on their own, or are a lovely accompaniment to the Pea and Mint Dip (see page 141) or the Chicken Liver Paté (see page 128).

Wastry Biscuits

Wastry (aka waste pastry) was an accidental invention when we had leftover pastry from making tarts. That day, we rolled wild garlic and salt into the surface of the pastry and baked it. What was born was beautiful buttery, salty, garlicky deliciousness and a zero-waste recipe that we snacked on all day. This recipe is very forgiving, sweet or savoury, so you can get inventive with toppings.

Prep: 15 minutes
Cook: 20 minutes

Makes 20 large crackers

Ingredients

200 g (7 oz) leftover pastry
20 g (¼ cup) herbs (such as rosemary)
Hearty pinch of flaky sea salt, for sprinkling on top
Flour of your choice, for dusting

Method

Preheat the oven to 200°C (180°C fan/400°F/Gas 6). Line a baking tray with greaseproof paper.

Roll out the pastry into a 3 mm (1/8 inch) thick sheet on a well floured surface. Wet the surface of the pastry with a pastry brush, then spread the herbs and salt over the surface (the water makes them stick). Sprinkle flour on top of the toppings and roll them into the wastry with a rolling pin.

Cut into 5 cm (2 inch) squares and arrange on the lined baking tray. Bake for 20 minutes, or until golden brown. Store in an airtight container for up to 3 weeks.

Jess's Fennel Buckwheat Grainola

As sweet tooths, we were originally not sold on the savoury granola idea but gastronomist extraordinaire Jess whipped up this recipe at a supper club. Served atop a beetroot, homemade créme fraîche and dill dish, the cayenne left a warm heat in your mouth and the fennel added a natural sweetness. It's also fab as a salad topping, sprinkled on labneh, soft cheeses and dips for mezze nights.

Prep: 15 minutes
Cook: 20–30 minutes

Makes 1 x 34 x 27 cm (13½ x 10½ inch) tray

Ingredients

200 g (7 oz) buckwheat groats
100 g (⅔ cup) steel-cut oats
120 g (scant 1 cup) pumpkin seeds
60 g (scant ½ cup) sesame seeds
1 tbsp fennel seeds
1 tsp sea salt
½ tsp cayenne pepper
½ tsp freshly ground black pepper
1 medium egg white
60 ml (¼ cup) oil

Method

Preheat the oven to 200°C (180°C fan/400°F/Gas 6). Line a baking tray with greaseproof paper.

Put all the ingredients in a bowl and mix together thoroughly. Spread the mix evenly over the tray and bake for 20–30 minutes until it is lightly crispy and golden.

Leave to cool on the tray, then transfer to an airtight container. This will last for 6 months.

* *NOTE: the spare egg yolk can be saved to make a richer topping for your moussaka on page 125. Rather than using 2 whole eggs, simply use 1 whole egg and 1 egg yolk.*

Fermented Ketchup

Ketchup is a great way to preserve a glut of tomatoes from the garden, or wonky ones you can sometimes pick up for a bargain at veg stalls. This recipe can be made with a variety of substitutes if you're happy to have a bit of flavour variation each time you make it. We have used lovage instead of celery because that's what we had from the garden, and sichuan peppercorns, as we had them in the larder, which brought quite a different flavour profile to other tomato sauces. We've also made it using kombucha vinegar (vinegar from kombucha that's gone too far).

Prep: 15 minutes
Cook: 1½ hours

Makes about 1 litre (4 cups)

Ingredients

1 tbsp oil
2 onions, chopped
Bunch of celery (or 5 good stalks of lovage, or a mix of both and a bit of parsley), chopped
3 garlic cloves, chopped
½ tsp toasted ground coriander
1 tsp ground allspice
1 cinnamon stick (or ½ tsp ground cinnamon)
½ tsp peppercorns
750 g–1 kg (1¾ –2¼ lb) tomatoes, chopped
100 ml (⅓ cup) raw apple cider vinegar
100 g (scant ½ cup) golden caster sugar
1 tbsp tomato purée
½ tsp hot sauce (or pinch of chilli powder)
Sea salt

Method

Heat the oil in a large saucepan over a medium heat and sauté the onions and celery or lovage for 5 minutes. Add the garlic, spices and seasoning and cook for another 2–3 minutess. Add the tomatoes, vinegar and sugar and bring to the boil. Reduce the heat down to a simmer, cover and continue to cook for about an hour, stirring occasionally to make sure it's not sticking.

Once it's looking thick and reduced, take the pan off the heat and blend with a stick blender until smooth. Sieve into a bowl to remove any large seeds or pieces of tomato skin. The ketchup will thicken a little when it cools, but if you want a stickier ketchup, return to the pan to reduce further.

For regular ketchup, keep in an airtight container or sterilized jars (see page 55) in the fridge for up to 3 months, or freeze in batches.

To ferment, pour into sterilized jars and store in a cool, dark place, checking and burping occasionally, keeping an eye on it to make sure it doesn't go too far. When it gets super bubbly and a bit tangy, refrigerate and use within a few weeks.

Vegan Mayo

We first learned to make this mayo at The Castle, our favourite urban rock climbing centre, veggie and vegan cafe and urban community growing space. We never looked back, and now always make our own vegan mayo instead of buying it, or making the classic raw egg-yolk based mayo, which pregnant women are told to avoid, meaning it isn't very inclusive for catering events.

Prep: 5 minutes

Makes 1 medium-sized jar (about 240 g/8½ oz)

Ingredients

50 ml (scant ¼ cup) soya milk, at room temperature
1 small garlic clove, crushed
1 tsp smooth mustard
3 tsp white wine vinegar
150 ml (⅔ cup) mix of vegetable, rapeseed, grapeseed and hemp oil
Sea salt and ground black pepper

Method

Use a handheld blender to blend the soya milk and garlic together in a bowl. Stir in the mustard and vinegar. Slowly add the oil in a continuous stream and keep blending. The mixture should start to thicken and quickly reach normal mayonnaise consistency. Store in the fridge in a sterilized 240 g (8½ oz) capacity jar with a lid (see page 55) for up to 2 weeks.

TIP:

- We use this on our vegan burgers, with fries, as a sandwich spread or on potato salads with gherkins and herbs. Add a few teaspoons of hot sauce or smoked paprika to spice it up, or more garlic to make it more like aioli.

Heritage Tomato and Nasturtium Side Salad

This side salad is great for adding a tang to any dish. Nasturtiums grow like weeds in many gardens. When we visited Berkeley, California, everybody's front lawn was adorned with nasturtium plants, and they spilled onto the pavement making great curb-side foraging. In addition to being beautiful, great for pollinators and for insect control in food growing, nasturtium flowers and leaves are delicious. We use the smaller leaves for a little less intense peppery flavour.

Prep: 10 minutes

Serves 2

Ingredients

150 g (5½ oz) heritage tomatoes, halved or quartered
10 small nasturtium leaves
2 tbsp oil
2 tsp apple cider vinegar or Blackberry, Soy and Basil Shrub (see page 68)
Pinch of smoked sea salt
3 nasturtium flowers (or any edible flowers), to garnish (optional)

Method

Place the tomatoes in a serving bowl with the nasturtium leaves. Mix the oil, vinegar and salt in a small bowl, then add to the tomatoes and toss together to allow the tomatoes to absorb the dressing for a few minutes. Garnish with nasturtium flowers or edible flowers if you have them.

Homemade Oat Milk

We use a lot of oat milk in recipes and for teas and coffees. However, there is nothing more rewarding than making your own, and when you can see how easy the process is, you may never go back. Plus, you can look after the pennies. After blitzing the oat mix you can add the leftover oats to cookies, savoury porridge (see page 91) or a face mask (see page 164).

Prep: 5 minutes
Cook: 5 minutes

Makes 800 ml (3¼ cups) oat milk

Ingredients

800 ml (3¼ cups) ice-cold water
200 g (11½ cups) rolled oats
Pinch of sea salt (optional)
½ tsp maple syrup (optional)

Method

Blitz the water and oats together for 10 seconds on high speed in a food processor or blender. Be careful not to over blend as you will end up with slimy oat milk.

Strain through a fine sieve. Do this twice if you want to make sure all the bits are removed. Don't use a cloth strainer and force the oats through as this will cause gritty oat milk. Add the salt and maple syrup, if using.

Store in the fridge in an airtight bottle or container for up to 5 days. Shake before using. The milk is best suited to porridge or cereal as it can separate in hot drinks like coffee.

Sourdough Starter

There are plenty of very informative bread books and tutorials out there on the science of sourdough but here's our simple guide to starting your own sourdough starter. We use this to make Rye Levain Crackers (page 142), Buckwheat Waffles (page 119) and Georgia's Leftover Grain Bread (page 83).

Prep: 5 minutes

Makes 500 g (1 lb 2 oz) starter

Ingredients

200 g (1¼ cups), plus 5 tbsp wholemeal or bread flour

Method

To make your own sourdough starter, mix a tablespoon of any wholemeal or bread flour with a tablespoon of water in a jar or tub of at least 500 g (1 lb 2 oz) capacity.

Leave out on the work surface, or somewhere you'll remember to 'feed' it. Feed with a tablespoon of water and a tablespoon of flour every day for 4 days. It should start to look bubbly and active, and smell a bit funky, fermented and sour – this means you've successfully managed to form a colony of bread yeasts.

On day 5 add 200 g (1¼ cups) flour and 200 ml (generous ¾ cup) water, so that you now have about 500 g (1 lb 2 oz) total. Leave for a few hours until it is bubbling and active, then you can scoop out 480 g (1 lb 1 oz) to mix with the seeds for the seeded sourdough on page 83.

Clean down

The clean down is something often slightly dreaded in the kitchen, however, we try to make this as pain free as possible. We have the most fun in so many of our events when cleaning down, eating leftovers and chatting rubbish. It has made for many fun bonding sessions, which is what cooking is all about.

The long-term effects of regular exposure to some of the chemicals in cleaning products is not fully understood, as they have only been around for a relatively short period of time. Volatile organic compounds (VOCs) in some cleaning products may even be a source of irritation to our lungs and may increase the risk of developing allergies or asthma, according to The British Lung Foundation. Environmentally, there are risks associated with the cumulative effects of many of these chemicals that leach into our waterways and soils, as well as the high carbon footprint of their production.

As well as the issues with chemicals, cleaning down opens a can of worms around disposable cleaning materials and throwing away bits of uneaten or inedible food. We don't want cleaning our kitchens to cost the earth, and we hate waste! Here are some helpful hints and tips on how to create the most eco-friendly clean down, and what to do with these so-called wastes.

Eco cleaning products

Reducing, refills, bulk-buying

A great way to reduce the carbon and water footprint and plastic waste associated with cleaning products is to reduce how many products you buy in the first place. By consolidating your cleaning cupboard you'll also save space and money. If you think you'll only use some specific kitchen cleaning product once, or very occasionally, it might be worth finding an alternative.

Refill shops, unfortunately, aren't widespread yet. But they are popping up in more cities and towns all over the world, and many branches of food stores even have a refill section, especially health-food stores and smaller supermarkets. Refilling bottles can help to reduce how much new plastic comes through your kitchen, but buying refillable cleaning products might also be a way to find more environmentally friendly cleaning products, as they are usually more suited to sensitive skin, aren't tested on animals and are those with less harmful chemicals.

Buying concentrated cleaning products may seem expensive, but considering how much further they stretch, this can be both a more economically sustainable option as well as more environmentally sustainable. The majority of any liquid cleaning product is going to be water.

We've used some simple homemade cleaning products for a while, reducing the need to purchase of new bottles of cleaners.

Simple all-purpose citrus cleaner

White vinegar is a great cleaning aid. It can dissolve rust and dirt. It also destroys many bacterias and fungi, which in turn helps to remove odour. Teaming it with your leftover citrus rinds makes a great-smelling surface cleaner. Be mindful not to use it on anything you have used bleach on (as it can cause fumes) and always store undiluted vinegar in a glass container.

Collect lemon, grapefruit, lime or orange rinds in a 500 g (1 lb 2 oz) jar. When you have filled the jar, submerge them in 500 ml (2 cups) white vinegar and store in the fridge to prevent the citrus from moulding. Leave to steep for a week.

Strain the liquid through a fine sieve into a 1 litre (1 quart) spray bottle using a funnel, measuring how much goes through. Add the same quantity of water to the mixture and shake well. Your spray is then ready to use.

Soda crystals

Soda crystals have long been used as an all-purpose cleaner. We've avoided buying heavy-duty chemical cleaners by leaving pans to soak with soda crystals – a top tip from Sadhbh's mum. Soda crystals leave no odour or taste on items that interact with food, like chopping boards.

To make different concentrations of soda crystal solution for cleaning, packets include the following guides:

Strong: mix 200 g (1 cup) with 500 ml (17 fl oz) hot water
Regular: mix 100 g (½cup) with 500 ml (17 fl oz) hot water
Mild: mix 20 g (1 tablespoon) with 500 ml (17 fl oz) hot water

A soda crystal solution, combined with a bit of elbow grease, is effective for:

- cleaning chopping boards
- unblocking sinks and drains
- cleaning washing machines
- cleaning dishwashers
- cleaning clothes (it is the main component of most powder washing detergents)
- cleaning extractor fan filters
- cleaning ovens and hobs
- restoring silverware
- cleaning stained chinaware
- using as a floor cleaner
- cleaning tiles and grouting
- softening water, reducing how much clothes washing detergent you need.

Soda crystals are not the same as baking soda or caustic soda, although they are similar, and they should never be mixed with bleach.

Heavy-duty hob and oven cleaner

These are two places where mess just seems to stick like glue. Abi Aspen spent many nights growing up baking cakes with her comrade in arms Ali (aka Dad) acting KP, and he would spend most of the evening just scrubbing the hob! If only we have known back then that when you team vinegar and bicarbonate of soda, the acid-base reaction causes a fizz, and helps loosen those tough stains.

To make your own cleaner, place 4 tablespoons bicarbonate of soda in a bowl, then add a little warm water to make a paste. Scrub your hob with the mix to remove stains. For the oven, decant 100 ml (scant ½ cup) white vinegar into a small spray bottle. Apply the paste, spray with white vinegar and leave it to sizzle for 15 minutes, then scrub to remove stains. If stains inside the oven are very stubborn, heat a large tray of water inside to produce steam, then scrub with your bicarb paste.

Disinfecting countertop spray

We used to use this homemade spray in the laboratory for disinfecting. It's a potent number, but can even kill a number of viruses. We prefer ethanol over isopropyl alcohol as that is crude-oil derived. Make sure to store any remaining 100 per cent ethanol in a safely secured cupboard out of direct sunlight.

While wearing gloves, mix together 150 ml (2/3 cup) filtered water and 350 ml (1½ cups) 100 per cent ethanol in a HDPE or glass bottle and store in a cool, dry place out of sunlight.

Kitchen cleaning aids

We try to avoid plastic scrubbers, sponges and cloths where possible due to their extremely long decomposition time, and use of non-renewable resources. But don't worry, there are many alternatives.

Paper towel alternatives

There are a growing range of paper towel choices on the market that make environmental sustainability claims. Often to be found in health-food stores, these include those made from bamboo and recycled materials, and many come in paper packaging and compostable plastic-like wrapping. However, supermarket recycled or 'eco' kitchen towel still comes in single-use plastic packaging. You can even try making your own by following the steps on page 161.

There are a handful of semi-reusable kitchen towels that are designed to be used for up to a week, being rinsed out before being disposed of.

Reusable cloth napkins/serviettes are of course an option for sticky fingers. Sustainability and environmentally focused online stores sell cloth napkins in the likes of organic cotton, linen and reused material. A colleague in the London food sustainability scene even started a company upcycling saris into napkins and table runners (www.ecofetes).

There are also kitchen cloths for wiping up spills. Even the oiliest mess can be washed out of a kitchen cloth with hot water and soap. Kitchen cloths can be put in a hot wash or boiled on the hob to sterilize them.

Sponges and scourers

We look for biodegradable and compostable fibres for our scouring pads and sponges. Readily available options for sponges include coconut fibres, cellulose sponges, hemp, loofa, walnut shells and sisal. You can purchase these at local health-food shops or online. Be mindful though, some products contain plastic or recycled plastic, rendering them possibly recyclable but not compostable.

Of course, we still have to consider and research the growing and manufacturing process, but at least the source material is renewable and many of these products are made out of byproducts from other industries.

If you find natural cleaning scourers aren't always cutting the mustard, literally, then of course have a more abrasive plastic or metal scourer waiting in the wings for those baked-on pans. Even reducing how frequently we go through non-renewable and non-recyclable items is a start. And you might find an eco-scrubber does the job nine times out of ten.

Brushes

With brushes we tend to look for FSC-certified wooden handles from fast-growing trees (such as beechwood), bamboo (which is technically a grass not a tree so cannot be FSC certified) or reclaimed wood. Make sure the bristles are also biodegradable or compostable (like sisal), as this is the part you will have to exchange at the end of its life. Some brushes come with replaceable heads, which is also a great way to reduce overall materials consumption.

Cloths

See our materials guide on page 43 for some of our favourite ecologically oriented materials.

If jay clothes are all you have available, you can machine or hand wash and reuse them to make the most of their life span. We also readily use old face flannels or scraps of old clothes, just pop them in the washing machine or wash by hand (you might want to boil them to sterilize the material), and they'll be good as new.

Mops

We follow similar guidelines when purchasing mops as we do when purchasing brushes. Mops were traditionally comprised of a wooden handle and cotton strings, which are niftily decomposable, eventually. These are also called string mops, and their durable cotton fibres are suitable for heavy-duty floor cleaning. They do require a wringing bucket, but a sturdy one will last a lifetime.

Cheaper plastic mops are more prone to breaking and needing replacing, and of course are not biodegradable.

Flat-headed, replaceable-cloth covered mops are a popular alternative for floor cleaning, and can be washed hundreds of times. A popular style is the microfibre E-Cloth mop brand. It's also easy to make replacement cloth covers with household material scraps, like towels.

Vacuum cleaners

Well-known vacuum cleaner manufacturers are making inroads to improve the sustainability of their appliances but beware of misleading claims and greenwash.

We've avoided vacuum cleaner bags for years with washable, reusable pneumatic cloth vacuum dust bags. Regularly emptying and ensuring good maintenance of your vacuum cleaner will extend its life. Look for local repair hubs or repair options to extend the life of all appliances, where possible.

Make Your Own Reuseable Kitchen Roll

Some online sustainability-focused kitchen stores sell reusable kitchen rolls as an alternative to paper towels, and we tried one tutorial for making our own. They can be stacked where you would usually keep kitchen roll, or if you have a roll holder then wrap this reusable replacement around that, fastening with snap fasteners or velcro.

Materials

20–30 fabric scraps, cut into squares of 28 cm (11 inches)
Sewing machine (or needle and thread if you're happy to hand sew) or hemming tape as an alternative
40–60 metal snap fasteners, or velcro (optional)
Scissors
Iron
Kitchen roll holder (optional)

Method

Start by ironing a 1 cm (½ inch) hem around the edge of your squares. Hem the edges of the kitchen towel squares by sewing or fastening with iron-on hemming tape. Attach velcro or snap fasteners to one edge of each square, so they can be attached. Attach each square and roll up, just like a kitchen roll, rolling around a kitchen roll poll if you wish.

Pull off a square as you need, for mopping up any spills. Once each square is used, put it in your laundry basket and wash as usual with other linens, towels and cloths, or rinse out in hot soapy water and allow to air dry before reattaching to your kitchen roll.

TIPS:

- You could also use old napkins to make this homemade kitchen roll, just cut them to the size listed above.
- Try backing the squares with equal-sized squares of towelling for extra absorbency.

Using waste

We met over a mutual dislike of food waste while cooking a dinner using food destined for the bin, in the knowledge that over one-third of all food grown worldwide is wasted. 70 per cent of UK food waste comes from within people's homes and is made up of: spoilage, offcuts, mis-purchases and packaging. Spoilage might be those carrots you forgot at the back of the cupboard, forgetting leftovers in the fridge or storing your bread wrong. We avoid this by making smaller shops, especially of perishables, and rotating our stock of supplies.

'Offcuts' is a word up for debate; many of our recipes include parts of plants and animals that you might not think to use or see as a 'waste'. This is a passion of ours, showing everyone how delicious things people think are inedible are! Leek tops, carrot skins, offal, whey sauce and bean juice are just a few you will find in this book. However, we do understand that for the general population this can sometimes be difficult. Some 'offcuts' can be easily used but we are not going to lie, some carrot top-pesto just isn't the tastiest, beetroot leaves get sad really quickly, and pumpkin seeds can be really tough or chewy when toasted. Sometimes these offcuts are just as useful being turned into compost in your garden, or if that isn't an option, food waste collection for anaerobic digestion (contact your local council if you don't have food waste collection. If you badger them enough they will often sort one out). Don't beat yourself up too much about this, or force yourself to make and eat something that is really unpleasant. Eating should be pleasurable. There is still so much more of the plant and animal than we realize that we can eat that does taste delicious.

Mis-purchasing is an easy thing to rectify. We understand it sometimes happens; you buy a tin thinking it's something and it's actually something else. This is an opportunity for a new recipe, however. If it's something you don't like and your friends don't want, the joy of global communication means there are many apps to sell or give away unwanted purchases. In the UK, Olio is a good place to start.

Packaging is a very tough subject. As a rule of thumb, we try to pick unpackaged goods. If this isn't possible we look for compostable (if you have access to compost) or widely recyclable packaging. We try not to buy single-use plastics (although we use other people's cast offs) but often there is a choice between organic packaged and non-organic loose. This poses an ethical dilemma, which you can make your own choice on (as yet we have found no scientific papers comparing the environmental impact of both), but we often choose unpackaged over packaged. Essentially, recycling still uses energy and single-use plastic of course ends up in landfill, being incinerated halfway around the world, and even makes its way into our oceans.

Oat groat hand and face mask

This is a great for a creamy, softening mask for those weather-beaten hands and dry faces. If you don't feel like cooking, it also has the added bonus of using up the leftover groats from making oat milk on page 150. We love using it after long days on the farm or hiking when we want our skin feeling soft and hydrated, but without putting on a mask with a long list of ingredients that we can't pronounce.

Mix 200 g (7 oz) leftover oat milk groats with 1 teaspoon honey in a tupperware. Load your fingertips with the mixture and apply to your face or hands. Put a lid on the tupperware and store it in the fridge for up to a week. Sit in a warm bath for 15 minutes until the mask has hardened, then wash off with warm water for silky soft skin.

Spent Coffee Grounds

Many of us just discard our used coffee grounds, however there are a few neat tricks to stop them being thrown in the trash, so you can get a second use out of this luxury food import. Here are a few ideas:

- Toss whole beetroot in a couple of tablespoons of oil, then roll in a coffee and salt mix made up of 100 g (3½ oz) coffee with ½ tsp salt. Place the beets in either a small, deep baking dish and cover them in more coffee grounds or wrap in tin foil and pour in more coffee grounds and seal the top. Roast in the oven at 200°C (180°C fan/400°F/Gas 6) for 1 hour. Peel the skin, slice and serve with créme fraîche, dill and homemade savoury granola (see page 144).

- Mix 50 g (1¾ oz) coffee grounds with 100 g (scant ½ cup) demerara sugar and 100 ml (scant ½ cup) melted coconut oil for a homemade body scrub.
- If you're feeling really adventurous you can cut the top off an old plastic bottle, layer the inside with mushroom spawn, corrugated cardboard and coffee grounds and grow mushrooms. For more details, visit https://grocycle.com/growing-mushrooms-in-coffee-grounds/
- Compost them!

Composting

One of the best things you can do with your food waste (aside from eating it) is composting, which creates food for plants from your waste food. Composting is important as it diverts any food waste that may go to landfill and produce harmful gases like methane.

Compost is measured by its two key components: carbon (C) and nitrogen (N). You are looking for a C:N ratio of about 30:1 for a good compost heap. Too much carbon and the pile will be slow to decompose; too much nitrogen and the pile will get too hot, wet and slimy and will break down using anaerobic bacteria and smell.

You don't have to get out your calculator to start a compost pile, just make sure for every bucket of food waste you use, add about three buckets of shredded cardboard or newspaper. Nitrogen-rich additions are usually more dense than carbon-rich ones. And remember, it's not just waste from your home you can use to make compost, garden waste and grass clippings can be nitrogen-rich additions, too.

Some examples of carbon to nitrogen (C:N) ratios are:

High carbon:
Shredded cardboard 350:1

High nitrogen:
Food scraps 12:1
Coffee grounds 20:1

Avoid foods like dairy, meat and household pet poop as these will attract unwanted guests to your pile, and they can also harbour pathogens that your compost pile isn't hot enough to kill off. Additionally, compost weeds pulled from your garden in a separate compost pile, as you don't want them to contaminate your compost.

Your compost heap can be just a mound on the ground, a ready-made or homemade container, or a tumbling variety. Just make sure you can cover the heap to prevent it from getting too wet, and prevent it from attracting larger critters who will want a chomp. Covering a heap should help keep moisture in through the substances you add to it, as you don't want it too dry either. We like compost heaps that touch soil if possible as they allow worms and other microorganisms to enter. Your compost heap will heat up, then cool, then microorganisms will enter and decrease the quantity of compost but increase its quality, then finally the compost will be ready. It might be worth having multiple compost heaps, so while one is the size you want and doing its magic, you can add to another.

We don't turn our piles as they are small, we just alternate between carbon and nitrogen layers. However, if your pile is large, it might be worth trying to turn it once. Composting isn't a quick process though, so get a cup of tea and chill out for six to twelve months!

Seeking a sustainable community

Since we started working in sustainable food systems and making them our raison d'etre, there has been an explosion in organizations, apps, networks and products facilitating improved food and kitchen sustainability. It's so heartening to see how much more often the link is being made between agriculture, food systems and global environmental change. For a long time it felt like food and food systems thinking were a separate discipline to environmental science and climate change, but a greater global shift towards understanding the transdisciplinary nature of both food systems and climate change means they are more often considered as interdependent.

Food waste was one of the hooks that got us, and many others, interested in food's environmental impact, considering how much food gets wasted, and how it can be better distributed to avoid this.

Community

We were put in touch with each other by Tristram Stuart, the founder of food waste-fighting and sustainable food systems advocacy charity, Feedback Global. Feedback is a great place to start if you're looking for a local network and people to connect with over food sustainability issues. They also have a global network. Likewise, The Real Junk Food Project has hubs and cafes throughout the UK, and has inspired pay-as-you-feel projects redistributing surplus food in all corners of the world.

Putting on our first food-surplus event, we also worked with Plan Zheroes, This is Rubbish, FareShare, City Harvest and FoodCycle. Since then,The Felix Project has also joined the ranks of well-established food-waste fighters.

Being active in food waste is just one way to make waves in a sustainable food system, however there are many more. Most of these organizations are charities and take volunteers, offer internships or even a variety of job openings. If you want to work at a more systematic level, tacking the root cause of a broken food system, there are great places to look for more information and opportunities below.

ORGANIZATIONS TRANSFORMING FOOD SYSTEMS

Organizations	Restaurants	Farming and trading	Apps/online
City Harvest (USA)	Blue Hill (USA)	Agricology	Farmdrop
Fareshare	Chez Panisse (USA)	Anson Mills (USA)	Food Sharing (Germany)
The Felix Project	Edible Beats (USA)	Better Food Traders – Growing Communities	The Good Shopping Guide
Food Ethics Council	Inver	Brooklyn Grange (USA)	Natoora (UK, France, Italy, USA)
FoodPrint (USA)	Lille Grocery (Denmark)	Coombeshead Farm	Olio
Foodsave	Noma (Denmark)	Hodmedods	Roots to Work
Food Tank (USA)	Rest (Norway)	Kiss the Ground (USA)	Seafood Watch (USA)
The Jeremy Coller Foundation	Silo	Land in Our Names	Too Good to Go
Love Food Hate Waste	Small Food Bakery	The Land Institute (USA)	Zoe
Nosh	Wakelyns Bakery	London Farmers' Markets	
The Real Junk Food Project		Soul Fire Farm (USA)	
Stop Wasting Food (Denmark)		Stone Barns (USA)	
WRAP			

Getting involved in sustainable food

We often get asked by people how they can get involved in food sustainability, and for sustainability hacks for their kitchens. Hopefully this book answers some of the latter, but for the former, we've compiled some suggestions.

WWOOF and Workaway

World Wide Opportunities on Organic Farms (WWOOF), and the Workaway platform are two popular ways of finding opportunities all over the world to learn more about food, agriculture and sustainability-focused work and living. Our friend, author and food writer Malou Herkes, amassed countless kitchen skills, food and growing experiences and sustainability hacks through WWOOFing, slow travel and homestays. She chronicles this in The Forgotten Pantry blog.

Urban farming and urban gardens

As we've mentioned, we first connected in an urban growing space in London, the Skip Garden. We've found that even when living in urban areas we've been able to connect to food growing, nature and other people with interests in more sustainable living through seeking out these spaces. In London these can be found through Capital Growth, a platform by Sustain. When we travel we seek these spaces out, from a tour of Brooklyn Grange in New York, to volunteering with a new urban growing space in Bali following a quick search on Facebook when we were in Indonesia. Despite the multitude of negative repercussions of the Covid-19 pandemic, many of which are ongoing and will long disrupt food supply, there were some positive local food system changes. Urban green spaces and urban food growing were highlighted as being of tantamount importance to communities, for both mental and physical wellbeing, as well as improved food self-sufficiency in many cases. Signing up for your local allotment scheme is another great way to get started with growing your own food in an urban environment, and finding your green fingers.

Commercial kitchens and working farms

Over the years we've met a lot of people looking to make their passion their job – to work with food, to open their own cafe or restaurant, to build their own farm or start their own food-sustainability project.

If it's kitchen skills and chef life you are wanting to dip your toe into, then even a day's volunteering here and there, or a stage, as it's called in commercial kitchens, could be a good starting point. Inquire at your favourite sustainability-focused supper club, restaurant or cafe.

You can also do this with farming, and it doesn't have to be as structured as WWOOFing. Some places you can contact and work with just for a day or two, or stay longer if that works. We have spent time at Stone Barns in the US, Weston Park no-till farm in the UK and at Kornby Mølle in Denmark. Just reach out, you never know what you might find.

Further reading & resources

This book is a starter guide on how to make sustainability-focused decisions in your kitchen, however, there is so much more we can all do. Our kitchens start from raw resources: fabrics, materials and food. This means researching and supporting those who produce those resources well. We have named many of our favourites throughout the book, but have listed some more resources below for you to do some more digging and seed sowing.

- Chef's Manifesto
- Eating Better
- Eat Forum
- Feedback Global
- Food and Agriculture Organization of the United Nations (FAO)
- The Food Ethics Council
- The Food Foundation
- Greenpeace
- SDG2 Advocacy Hub – Chef's Manifesto
- Sustain
- Sustainable Development Goals
- The Sustainable Food Trust
- The Sustainable Restaurant Association
- World Wide Fund for Nature (WWF)

Events, talks and books

We love a good panel discussion or guest speaker at an event. These resources help broaden your knowledge of something someone else knows a lot about. Regenerative agriculture and sustainability are hot topics. Look out for publications and talks by some great thinkers such as Alan Dangour, Marco Springmann, Helen Browning, Gabe Brown, Rosemary Green, Dan Barber, Vandana Shiva, Naomi Klein, René Redzepi and Patagonia clothing for a wide coverage of all things sustainability focused.

Some background reading

This is a list of titles to get your teeth into if you are looking to discover more on the food systems, agriculture and environment nexus:

- **Rachel Carson** *Silent Spring*
- **Sarah Bridle** *Food and Climate Change Without the Hot Air: Change Your Diet: the Easiest Way to Help Save the Planet*
- **Tim Lang** *Sustainable Diets: How Ecological Nutrition Can Transform Consumption and the Food Sytem Feeding Britain: Our Food Problems and How to Fix Them.*
- **Michael Pollan** *The Omnivore's Dilemma: The Search for a Perfect Meal in a Fast-Food World*
- **James Rebanks** *English Pastoral: An Inheritance*
- **Allan Savory** *Holistic Management: A Commonsense Revolution to Restore Our Environment*
- **Vandana Shiva** *Soil Not Oil: Environmental Justice in an Age of Climate Crisis*
- **Tim Spector** *Spoon-fed: Why Almost Everything We've Been Told About Food is Wrong*
- **Carolyn Steel** *Hungry City: How Food Shapes Our Lives Sitopia: How Food Can Save the World*
- **Dan Saladino** *Eating to Extinction: The World's Rarest Foods and Why We Need to Save Them*
- **Tristram Stuart** *Waste: Uncovering the Global Food Scandal*

About us

Together we are The Sustainable Food Story, a collective who run ecology-inspired dinners and experiences. We met at the Skip Garden in 2017, where Sadhbh worked with charity Global Generation. Abi Aspen needed a launch venue for cooperative farming project #ourfield, and from that point on we have been inseparable – a food campaigning duo. We have always been very keen on not just talking about the problems in the food system but 'walking-the-talk' – getting involved in a spectrum of work from farming to cheffing, nutrition to youth work.

Abi Aspen is a farmer, scientist and environmental entrepreneur. With a MEng in Chemical Engineering, she started her food systems career as a cellular agricultural scientist before swapping lab-grown meat for fields of wheat in 2017, the same year TSFS was born. Upon leaving the lab she volunteered for a few months at Blue Hill and the Stone Barns Center, mostly in the fields but also in the bakery, before returning back to the UK. She now works as a part-time farmer at Duchess Farms – a regenerative collaborative farm in Hertfordshire. After crowdfunding for a milling system in 2020, she now runs milling operations at Duchess. In her most recent endeavour to bridge the world of biotech and farming, she has recently co-founded a start-up as CTO, upcycling agricultural by-products and forgotten crops into next generation proteins. In the down season she sometimes pops up as the guest editor of Sustain's *Jellied Eel* magazine.

Sadhbh is an eco-chef, nutritionist and sustainable food activist. She has an MA in Sustainable Development from the University of St Andrews, and an MSc in Nutrition for Global Health from the London School of Hygiene and Tropical Medicine. Her initiation into environmental campaigning was working for Greenpeace on the Save the Arctic campaign. Since 2014, she has worked for a range of environmental, youth and food sustainability NGOs, including Global Generation, Julie's Bicycle and Eat Club. In pursuit of her passion for challenges and learning, she most recently worked in polar food operations for the British Antarctic Survey in Antarctica.

When not knee-deep in the food system you will find us climbing rocks, swimming outdoors or eating homemade cake.

Index

First published in 2022 by White Lion Publishing
an imprint of The Quarto Group.
The Old Brewery, 6 Blundell Street
London, N7 9BH,
United Kingdom
T (0)20 7700 6700
www.QuartoKnows.com

A catalogue record for this book is available from the British Library.

ISBN 978-0-7112-6576-9
Ebook ISBN 978-0-7112-6577-6

10 9 8 7 6 5 4 3 2 1

Art direction and design: Rachel Cross
Commissioning editor: Melissa Hookway
Editor: Charlotte Frost
Food and props styling: Polly Webb-Wilson
Publisher: Jessica Axe

Printed in China

THANK YOU

This book wouldn't have been possible without so many people. We'd like to thank our parents, friends, colleagues and partners. We have peppered this book with references to some of the amazing people, mostly badass female chefs, whom we have had the great fortune to work with and learn from over the years in so many inspiring sustainability-focused kitchens and projects. Thanks to those who took us in when we knew so little, helped us learn, make mistakes and keep going, making space for us to grow. We have learned so much, and run The Sustainable Food Story with a mantra of kindness, empathy and love, which we have received and hopefully spread to others.

An extra thank you to Jenny, for allowing us to use her wonderful home as a studio, but also as the test kitchen for many of our dishes, our early days of forming The Sustainable Food Story, and her wisdom for many of the sustainable kitchen techniques and craft guides.

The authors and the publisher would like to thank The Arty Vegan, Chapel Fruit and Veg, Isle of Wight Tomatoes, Nigel's Lettuce and Lovage and all of the other business owners who kindly allowed us to shoot at their stalls and shops.

Thank you to to Oscar and all of the Harding family at Duchess Farms for their hospitality, which resulted in the beautiful outdoor shots in the book.